D0344901

ABOUT THIS BOOK

The little-publicized hazards of electric power transmission, and the short-sighted, materialistic policies of the industry behind it, are laid bare in this timely book. For this, the paper edition, Louise Young has added appendices containing practical advice on *"what you can do about transmission lines"* and an epilogue describing the events that have transpired since this controversial book brought home to Americans the dangers inherent in the construction of high-voltage transmission lines. Studies conducted in Russia conclude that long-term exposure to fields of this strength causes cardiovascular changes and damage to the nervous system.

"An admirable story of an individual fighting the corporate giants and a contribution to a neglected aspect of the energy scene." –*World Environment Newsletter*

"A beautifully executed personal statement with profound meaning for us all. . . . A great little book, with some lovely photographs and an extensive bibliography." –*Library Journal*

"With both the scientist's expertise and the writer's deep concern for the people involved, [Louise Young] explains both the aesthetic and the practical problems of the present system." –*The Science of the Total Environment*

"A careful, well-argued case for considering alternative means for generating and transmitting electrical power." –*The Annals of The American Academy of Political and Social Science*

"An intelligent combination of politics and economics that also manages to portray the human beings involved as more than one-dimensional." –*Washington Monthly*

"An informed, convincing, and well-written book with implications that reach far beyond the immediate situation it describes." –Paul B. Sears, *Growth and Change*

Power Over People has been awarded the Special Certificate of Merit by the Friends of American Writers.

Louise Young is a physicist, Director of the Open Lands Project in Chicago, and a member of the Executive Council of the Lake Michigan Federation. From 1962 to 1973 she was Science Editor of the American Foundation of Continuing Education. She is the editor of *Exploring the Universe, The Mystery of Matter, Population in Perspective, Evolution of Man,* and (with W. Trainor) *Science and Public Policy.*

Power Over People

Power Over People

Louise B. Young

OXFORD UNIVERSITY PRESS
London Oxford New York

OXFORD UNIVERSITY PRESS

London Oxford New York
Glasgow Toronto Melbourne Wellington
Cape Town Ibadan Nairobi Dar es Salaam Lusaka Addis Ababa
Delhi Bombay Calcutta Madras Karachi Lahore Dacca
Kuala Lumpur Singapore Hong Kong Tokyo

Library of Congress Catalogue Card Number: 72-91020

First published by Oxford University Press, New York, 1973
First issued as an Oxford University Press paperback, 1974
Printed in the United States of America

To my son, my two daughters, and the other members of their generation who understand that the pursuit of profit at the expense of the beauty and integrity of our environment is impoverishing us all.

Foreword

The beauty of this book is that its author, Louise Young, has given us a graphic, saddening description of what happens when the so-called necessities of a giant electric power company conflict with the desires of countrymen and their community. As one would expect in this era of surging "progress," the company wins and the people lose. However, Louise Young was there, and her jarring account of the classical conflict that arises when government and industry give greater importance to "progress" than to the living values of the people is the story of *Power Over People.*

The setting of this conflict is an Ohio village—population 104. It could instead be any part of rural America which is richly endowed with natural beauty—and citizens who want to live out their lives in an unsullied environment. In the 1960's, Mrs. Young tells us, this town possessed essentially the same character it had had fifty years earlier. Her homeland was off the beaten track, and it had accepted a few of the benefits of modern technology without abandoning the closeknit fiber on which the community's life was built.

So in 1969, when the people of Laurel learned that one of the world's largest electric transmission lines was planned to plow through their backyards, they banded together to oppose it.

With misplaced trust, they sought at first the relief of the law and the regulatory agencies that had been created to protect the "public interest." To their surprise, they found that these indifferent agencies had no authority to respond to their complaints and that the law had given the electric power company almost unlimited power (we call it "eminent domain") to alter the landscape at will. As the citizens of Laurel struggled to engage the company in battle, it finally became clear that their interests were really being sacrificed to the electric power "needs" of two growing metropolises hundreds of miles away.

Ironically, the power to be transmitted over the new line was destined for Detroit and Chicago, cities whose inhabitants would not tolerate the construction of additional coal-burning plants within their boundaries. So instead, the power company decided to export its pollution to the countryside and to build a huge mine-mouth generating station and transmission lines to get the electricity to market. This is a common practice now, and one which in recent months has aroused bitter disputes in many parts of the United States, as environmentalists, who believe that cities have no right to turn the countryside into wastegrounds in order to accommodate further urban sprawl, fight against the further degradation of unspoiled America.

In the end, the citizens of Laurel made an important impact but failed to save their own community. One of America's technological feats will soon overpower their scenic landscape—a tribute to an industry that has been obsessed with the goal of "cheap power," with virtually no concern for anything else. The long-prized character of this town is doomed, although the traditional wisdom of government and industry decisionmakers tells us to accept this in the name of "progress" and other material benefits.

This is a disturbing book—as Mrs. Young intends it to be. Little is being done today by government and industry to prevent the destruction of more and more of the country's Laurels. By making the single-minded argument that the projected demand for electricity must be met, some industry spokesmen are working under the assumption that angry environmentalists are at the root of the escalating national energy crisis.

I agree with Mrs. Young that the nation is making a wholesale sacrifice of environmental values to serve the burgeoning demands of unprecedented exponential growth in energy consumption, some of which is inherently wasteful.

In my view, the answer lies in another direction. As a nation, we must adopt national energy policies predicated on conservation rather than on waste. We must open the doors of democratic debate to all concerned citizens, not let the energy industries bulldoze their critics by arguing that the lights will go out unless their ill-planned projects move forward forthwith. We must provide balance between the forces of industry and environment and curtail such sledge-hammer threats as the power of eminent domain. And, above all, we must learn to live with restraint; to accept nature's limits as the first principle of creating a livable human environment; and to support those who are fighting to slow down population growth and make it possible for us finally to concentrate fully on the quality of life for America's future.

Mrs. Young's book will help educate the public on the plight of ordinary citizens who challenge the wisdom of "semi-secret" government and industry decisionmaking. Her perspective is sharp as she writes with conviction and idealism:

> As we put into orbit satellites that beam messages around the earth in seconds and see on our television screens views of the earth that encompass three continents in a single shot, it is impressed upon us that our home is limited both in space and in time. Earthspace is precious. There will never be any more of it. It can only be stretched by learning to use it more wisely.

As the push for greater energy production continues to intensify, we can only hope that the nation adopts this principle and uses our land more wisely, and that the broader quality-of-life values that Mrs. Young writes of so well will supplant the single-minded alliance of technology and economics that has brought us to the environment-energy crisis of the 1970's.

Washington, D.C.
October 1972

Stewart L. Udall

Acknowledgments

This story is based upon events that occurred in southern Ohio between September 1969 and May 1972. Although I have used some literary license in describing scenes and personalities, these changes have been introduced only for the purpose of portraying more vividly the real indigenous characteristics of this part of the country. The episodes involving the power company describe actual experiences as reported to me by the local people. Names of individuals and of the two small villages have been altered to protect personal privacy.

In a very real sense, I am grateful for the events related here because they made me aware of the ultimate consequences of values that have been too easily accepted. Like most members of my generation, whose point of view was formed during depression years, I had grown up with the proposition that "progress" meant a steady march toward a higher level of economic security and would lead inevitably to an improvement in the quality of life. My experience with the power company brought dramatically to my attention the fallacies of that belief. By opening my eyes to the inexorable advance of destruction created in the name of "progress," it provided the motivation for this book.

The original research that went into the project was largely a family affair. My son, H. Peyton Young, Assistant Professor of Mathematics at City University of New York, contributed many hours of library research. My daughter Anne, who will receive her M.D. from

Johns Hopkins in 1973 and is working simultaneously for a Ph.D. in neuropharmacology, gave expert assistance on the medical and biological aspects of the problem. My younger daughter Isabel contributed many valuable suggestions and discovered relevant articles. My husband, Hobart P. Young, added his knowledge of chemistry and his understanding of the business world. In addition he spent many hours reading and criticizing the manuscript as it progressed. I am grateful for the encouragement and time so generously given.

Although final responsibility for the accuracy and validity of the text is mine, I am indebted to a number of scientists who read portions of the manuscript, provided technical information, or offered valuable suggestions. I especially want to thank Cheves Walling, Professor of Chemistry, University of Utah; Charles H. Baer, Associate Professor of Biology and Plant Ecologist, West Virginia University; Leon S. Dochinger, Plant Pathologist, U.S. Forest Service, and conservationists Richard S. McCutchen and Jeffery R. Short, Jr.

I also want to express my appreciation to Beverly Lawson, Phyllis Dubnick, Virginia Bauman, and Stephanie Golden for their capable assistance in putting the manuscript into finished form. And special thanks go to my editor, William C. Halpin, whose belief in the project helped me over several discouraging times when the work might have been abandoned.

Finally, I am deeply grateful for the inspiration provided by all the staunch people who stood together to fight the erection of the high-voltage line across their land, and to the many who have written or called me expressing their concern about similar construction projects throughout the country. I believe that the spirit and enlightened sense of values demonstrated by the people who speak out against this wanton destruction of our countryside will ultimately prevail. We will learn to subordinate the profit-motive of our materialistic society to the needs of the whole-earth system. By tending and preserving our superb natural legacy, we will actually pass on a much richer inheritance to the future.

Winnetka, Ill. L. B. Y.
October 1972

Contents

Power Over People

1

A Quiet Place

As you drive east from Kinnikinnick on Ohio Route 180, the countryside becomes gently rolling with varied vistas of rich farmland. You pass over a series of little roller coaster rises, then over a sharp crest, and you come suddenly upon a small village nestled in a hollow between two hills. A white sign announces the dimensions of the town:

LAUREL
unincorp
Settled in 1807
Population 104

About three dozen houses of assorted ages are strung out along the roadside, each with its plot of grass, its dahlias and nasturtiums, its line of laundry flapping out in back, and in front a porch with a chair or two for sitting out on a warm day. A steepled church stands at each end of the village and the center is marked by a pair of red gas pumps in front of the general store.

The size and personality of this village have not changed appreciably in the half-century that I have known it. Innovations have come and have been absorbed slowly and inconspicuously, improving its comforts without destroying its character. The brick one-room schoolhouse is no longer used for its original purpose. The children are bussed to a central school eight miles away, but the schoolhouse has other uses now. It is a gathering place for special occasions—4-H club meetings, fish fries, and strawberry socials. Almost every house in the village has a well, indoor plumbing, a telephone, television, and electricity. As a matter of fact, Laurel was one of the last communities in this part of the United States to receive electric service. For many years the nearest lines were five miles away, and the power company said that it was too expensive to extend the lines five miles to service this small village and outlying farms. As late as 1936 gas and kerosene were still used to light the homes here. It was not until 1937 under the Rural Electrification Administration that they were electrified.

Fortunately, however, the town has also been bypassed by most of the obscenities of progress. It has no neon lights, no billboards, no housing developments, no supermarkets. The general store is just one room large but it provides a rich assortment of shopping possibilities. There one can pick up the mail, collect the evening newspaper, and hear the local news, as well as buy frozen foods, hardware, nylons, fresh country eggs, and almost any staple grocery item. There are always baskets of fresh vegetables and fruits from the neighboring farms and several specialties for sale such as Pearley Jones's dark clover honey and the wonderful stuffed sausage made from the proprietor's old family recipe and sold here for the past hundred years.

All in all, Laurel is a very good place to live—clean and quiet with miles of beautiful unspoiled countryside right at its doorstep. To the south and east rise the foothills of the Appalachians. Here are 30,000 acres of state forests with picnic areas, camping grounds, and nature trails. Here are limestone formations carved by glaciers into spectacular deep grottos, box-canyons, room-size caves, and hidden waterfalls. Beyond the state parks lie miles of wooded country, extending nearly to the Ohio River, covered with forests of oak, maple, and pine. People who live nearby can enjoy the forests in all seasons. They can walk there in the autumn when the flaming crimson of sumac floods the little glens and valleys with color, or take a picnic supper to Tar Hollow in the spring when the park is starred with dogwood and carpeted with banks of blue myrtle.

On the other side of Laurel, to the north and east, the land falls gently away and then levels out into a wide bowl of very productive farmland. This valley was one of the original prairie lands discovered by the settlers who came out to Ohio in the eighteenth century. An unusual characteristic of the Great Lakes region, these expanses of grassland set in among the hardwood forests seemed especially remarkable to the early pioneers. Letters and reports sent back east contain descriptions such as these: "I could not help pausing frequently when I struck the first burr-oak opening I had ever seen, to admire its novel beauty. It looked more like a pear-orchard than anything else to which I assimilate it—the trees being somewhat of the shape and size of full-grown pear trees, and standing at regular intervals apart from each other on the firm level soil. . . ." "The prairies consisted of level stretches of country covered with sedge-grass, and dotted here and there with patches of scrubby burr-oak growing upon the highest points of land. The sedge-grass grew to an enormous height, sometimes sufficient to hide man and horse when traveling through it."

The pioneers soon discovered that these small stretches of open country were particularly desirable as farmland. Crops could be planted more easily there, without the labor of clearing forests. The deep sod was hard to break but the dark loamy soil uncovered by the plow was richer than the lighter colored soil of the forest land. The larger prairies in Illinois and farther west presented greater difficulties for the homesteader. They were windswept and more exposed to extremes of climate; there were no nearby forests to provide timber for houses; and the grasses did not grow as tall as the big bluestem and the Indian grass of the small eastern prairies.

Naturalists are undecided about the origin of these small eastern prairies. Why should one piece of land support prairie while another nearby is covered with maple and beech trees, with trillium and bird's foot violet in the shade below? Many answers have been suggested. Dense prairie sod, once established, is a poor seedbed for trees. Frequent fires favor grassland over forest. The roots of many trees need fungi in unions called mycorrhizae to function well; perhaps these fungi were absent from the grasslands. One of the most favored explanations is that these prairies were established during a dry period following the last glaciation. Grasses flourish and expand at the expense of forests in dry climates. In more recent times the climate of this region has become cooler and moister, allowing the forests to encroach on the prairies, so that the prairies have become small open expanses set among woods. If this theory is true, the eastern prairie represents a relic from very early days. The tall grasses which the pioneers found there had grown undisturbed for centuries and had built some of the world's most fertile soil.

The prairie north of Laurel was settled early. My ancestors came before 1800 and bought 700 acres of property from Isaac Dawson, a Revolutionary War hero who had acquired the farm by a land grant. The original property contained a large log barn and a two-story log house. The upper floor of this house was one enormous room which served as dance floor and meeting place for the other settlers and their families. The barn, which still stands, was built of hand-hewn solid timbers and fastened together with wooden pegs. According to local tradition, church services were held here before the Lutheran church was built in the village.

For seven generations the land on this farm has been cultivated conservatively, the crops rotated year after year, the thistles cut down by hand, and clover plowed back under the soil so that the cycle of productivity could begin again. With improved farming methods, the system has been updated. Rotation was changed from a four-year to a three-year cycle. Just very recently the prairie fields have been planted in continuous corn. Since the fodder is no longer removed and used as bedding for livestock, it can now be plowed back to provide organic humus. Manure from the feedlots is spread on the land and chemicals are added sparingly as needed. Year after year the land responds by putting forth larger crops.

From the farmhouse porch on a midsummer day you can look out across hundreds of acres of prairie land where row upon row of young corn converge toward a distant rim of blue hills. Light breezes stir the green leaves, turning the scene into a rippling sea of light and shadow. If it is a typical summer afternoon the loudest noise may be the drone of bees in the honeysuckle and the occasional deep clunk of a bull frog down by the pond.

This is deep country, where one feels totally immersed in a harmony of man with nature. It is not a breath-taking scene like the Grand Tetons or the Amalfi Coast but it is wonderfully peaceful and restoring. To all members of the family that has lived there for almost two centuries this view is the heart and soul of the farm. On the day before my father died, after a long and terrible illness, he asked to be carried over to the window so he could "look out across the fields once more." His wife and sister carried him; he weighed only seventy-nine pounds.

It is easy to write off this kind of affection and personal involvement as sentimentality; and there are many who feel that sentimentality has no place in the world today. These people should be cheered by the knowledge that there are fewer and fewer places on earth where a man can experience any creative partnership with a portion of nature. True, he can still go on a camping trip to Glacier Park, but that is not the same kind of experience. In two weeks it is not possible to participate in the intricate interrelationship of living things or to sense the deep inner rhythms of nature. But the tune of our times is set to another rhythm—a noisy accelerating rhythm which searches out every quiet place. Sentimentality is a luxury that cannot be tolerated; it might get in the way of progress.

It was September 1969, and in the central offices of the Ohio Power Company in Canton, a plan was approaching fruition. Several years of study and computation by Ohio Power and its parent company American Electric Power had resulted in the decision to build a large power plant near Cheshire on the Ohio River and to carry this new power diagonally across the state of Ohio to connect with lines leading into the Detroit industrial area and with Commonwealth Edison's lines just south of Chicago. Power could be sold at a good price to these vast industrial complexes and could be generated with maximum economy by using the cheap high-sulphur coal strip-mined near the Ohio River. The Chicago and Detroit areas, and many other parts of the country, were beginning to enforce regulations forbidding the use of high-sulphur coal without installation of expensive anti-pollution devices, but southern Ohio was not one of these areas. Economi-

Yoichi R. Okamoto

An accelerating rhythm that searches out every quiet place.

Eliot Porter

cally it appeared to be advantageous to produce the power for these cities on the Ohio River even though it meant transmitting the power all the way across the state.

Maximum economy in transmission was a key factor in this plan. Along the diagonal path of the line there were several places that had to be avoided. Large city areas like Columbus are expensive to cross, for lines must be put underground. Avoiding the city would require a bend in the line; this could be handled most conveniently at a town named Marysville. A straight-edge laid across the map from the chosen point south of Cheshire to Marysville intersected the little village of Laurel. It was a fortunate thing, executives of the power company observed, that the whole lower third of the line passed through rural areas and "backwoods" country. They knew from experience that they were less apt to encounter resistance from poor and uneducated owners. As one man put it, "It's easier to screw the hillbillies."

The plan had been a well-guarded secret. It was very important that no advance knowledge should alert people along the intended right-of-way; there must be no opportunity for community action opposing the erection of the line. But each piece of property had been photographed and plotted from the air. The route representing the greatest economy had been precisely determined and drawn up in detail. All that remained now was to notify those concerned and force as rapid a settlement as possible.

The agents worked in groups. Armed with their drawings and contracts, they fanned out across the countryside near Laurel. Adjoining farms were approached simultaneously in order to maximize the surprise element and to minimize the possibility of neighbors consulting with each other. Tempted by the payments offered in compensation—although these payments were actually very low—a number of farmers signed on that first day. Others held back, postponing the evil moment they knew to be inevitable. They were far-sighted enough to understand that the sum of money they would receive would be spent in a short time but the transmission line would still be there for their lifetimes, perhaps for their children's and grandchildren's. The ones who held back benefited financially in the long run. Several of the last ones to settle were paid seven times the amounts they had been offered initially.

An agent knocked on our door that day but found no one at home. By the time he did find us about two weeks later, we already knew the basic facts and thought, hopefully, that we had some reasonable compromises to suggest.

The agent, Mr. Jackson, was a small man with a brisk, decisive manner. He was anxious to complete this transaction as quickly as possible. It was already late in the day and he hoped to be back in Canton that evening. We led him out onto the porch and sat down facing the view of the prairie. Mr. Jackson spread his drawings on the table. The route that had been selected for the power line, he explained, cut diagonally across the widest part of our farm from the southeast to the northwest corner and passed directly between the house and the prairie fields. It would cross just to the north of the pond there, he said.

My eyes followed his pointing finger to where the soft line of willows ran down along the brook to the pond. The glow of late afternoon sunshine turned its still surface to a mirror of gold and highlighted the yellow leaves of the great sycamore tree. In its shade the cows stood with heavy udders, rhythmically switching the flies off their backs, waiting for the evening milking.

Yes, Mr. Jackson admitted, the sycamore tree would have to go. And the willows. The right-of-way would be 200 feet wide, and in this strip no trees or shrubs would be allowed to grow. Somewhere there—perhaps where the sycamore now stood—would be a tower one hundred and twenty-five feet tall. Two more towers would stand between the house and the fields.

How much would it cost, we asked, to put the line south of the house where it would not intercept the view. He smiled at the naïveté of this idea. "A billion dollars," he said, "would not move this line one foot!"

"This is no ordinary power line," he went on with obvious pride. "It will carry 765,000 volts, the highest voltage of any line in the world."

Perhaps he sensed then that he had revealed too much, for we were unable to obtain any more information from him about the design and construction of the line. In fact, as we discovered later, no power company officials would release any technical information concerning it. Our attempt to find answers to our questions led us, during the next two years, into university laboratories, technical libraries, and government offices. The facts we unearthed about the new extra-high-voltage lines are presented here to the public for the first time. The hazards associated with this "technological triumph," the short-sighted and materialistic policies of the industry that lies behind it, should be matters of deep concern to the American people.

2

Power Transmission Pollution

Everyone agrees that electric transmission lines are appallingly ugly, but suggesting that they may also be sources of audible and chemical pollution is as surprising as saying that the homely freckle-faced boy next door is guilty of rape and sodomy. Most electric transmission lines look innocent enough. No flickering fires play along the wires. No smoke pours off them. The air in their vicinity appears unaffected by their presence, and when we pass under them there is no sound or sensation of the millions of kilowatts that are passing directly over our heads.

This situation, however, is rapidly changing. Unsuspected by the vast majority of people, thousands of miles of transmission lines that will produce a wide spectrum of visual, audible, and chemical pollution effects are being rushed to completion throughout the United States. These lines are just the beginning of a new trend in power transmission that threatens to degrade millions of acres. Unsightly metal structures will stand as tall as twelve-story buildings on fertile farmlands and wooded hills across America.

Visual Pollution

Over the past century citizens of industrialized countries have become acclimated to the visual pollution of overhead lines. Acceptance has occurred by slow degrees in the same way that all forms of pollution become established: they begin with very small changes in our environment, so small that they seem to be totally insignificant, and increase by tiny increments day by day so that the changes are hardly noticeable. Lulled by long familiarity with the polluting factor, we hardly notice the steady but inexorable degradation of our environment until one day we wake up and realize that the lake is too polluted to swim in, the air is dangerous to breathe, and the fish are contaminated with DDT. By this time the damage has gone so far that to correct it requires vast expenditures of money and effort. It may necessitate the reorganization of an entire industry—such as the automotive industry—or the nationwide reconstruction of a public service such as the waste disposal system. Because the demand for the service or the product has escalated along with the source of pollution, crash programs must be undertaken in order to find an acceptable alternative before the damage can be arrested.

If we are serious about preserving our environment, problems must be recognized early, potential damage assessed realistically, and alternatives considered before any industry or public service has become very deeply committed with large capital outlay in machinery and equipment. We should beware especially of those who attempt to justify new factors known to be damaging by such phrases as: "The effect is hardly any more than . . ." or "It represents only a small percentage of the total to which we are exposed" or "It is so minor a change as to be negligible." Any change which is harmful and which, by its nature, may increase or be cumulative should not be permitted to become entrenched.

The acceptance of overhead lines began with the invention of the telephone. The public's enthusiasm for this remarkable new form of communication overcame any resistance that might have been felt against these ugly additions to our landscape. At any rate, telephone poles are relatively low, partly obscured by trees and buildings. We got used to them, and after a while we didn't notice them any more. As telephone service increased and was brought to rural communities, we became accustomed to seeing the lines stretch out along country roads, silhouetted against the countryside. In general, they followed streets and highways on public property and rarely crossed private property except to bring service to that person's home. It didn't occur to many people to object to this minor form of visual pollution.

Then, with the invention of electric lights, lines very similar in appearance to telephone lines brought this new magic into our homes. At first, the new lines followed the same paths as the telephone lines. Five wires were not *much* worse looking than two. The poles were a little higher but just a *small* amount. We got used to that change, too. Then, as the electric industry learned how to produce power at higher voltages, taller poles became necessary. This change also took place so gradually that we didn't really notice. Electrical engineers, however, knew that power could be transmitted more economically by running the lines directly from the source of power to the point of consumption. The newer electric lines began to be constructed across private property.

About this time, towns and suburban communities recognized that the tangle of wires and poles along the streets confined traffic patterns and impeded construction projects. Eventually some communities passed ordinances requiring that lines be placed underground. But underground transmission was more expensive to install; so, as soon as the town limits were passed, the electric and telephone lines emerged again into full view.

In the meantime, the electrical engineers were making rapid strides in designing higher voltage equipment. Great stress was placed on the development of this technology because electricity is transmitted more economically at higher voltages. The power delivered is the product of the voltage and the current. But the losses caused by heating along the line are least when the power is transmitted at high voltage and low current. Then, since high voltages are very dangerous, the transmission voltages are transformed to safer levels near the locations where the power is to be used. It is delivered to the customer at 110 and 220 volts.

There is, however, one disadvantage to high-voltage transmission. The lines that carry the power must be larger, or a different kind of loss called *corona discharge* begins to occur. In order to keep losses to a minimum, either the diameter of the conductors should increase as the voltage increases or the number of them should be increased. One consequence of the necessity for larger or more numerous conductors is the need to build stronger supporting systems. The poles of the lower voltage lines are not adequate for high-tension lines. Steel or aluminum towers usually support the conductors that carry high voltages for transmission over long distances.

During the late forties and early fifties these towers began to mushroom over the countryside, and the heavy lines swooped in long arcs against the landscape. There was no longer much resemblance between the relatively inconspicuous telephone lines and the high-tension electric lines, but the principle of acceptance had been established and there was no public outcry.

The towers carrying the 138,000-volt lines were just the beginning. By 1952, lines carrying 345,000 volts were being erected. These lines, of course, required much heavier conductors and more massive steel towers. In the 1960's several lines were energized at 500,000 volts and then, in 1968, construction was started on a network of lines designed to carry 765,000 volts. The towers for these lines are enormous metal structures, measuring 90 feet at the cross arms and 120 to 135 feet high. These are the towers that will stand like so many steel scarecrows between our house and the view of the valley. Where today we watch the swallows dip low over the fields in long graceful arcs, tomorrow we will see the swoop of heavy conductors outlined against the hills.

"You'll get used to them," a power company official told me. "You'll be surprised how quickly it will happen. After a little while you'll hardly notice them at all."

The really frightening thing is, of course, that he is right. This ability to look without seeing is part of the adaptation we are all making to a rapidly deteriorating environment. We look *around* billboards and *over* superhighways and *under* transmission lines and pretty soon we don't really *see* at all. In an effort to protect ourselves from the jarring impact of ugliness we are slowly becoming desensitized. It is alarming to think what this by-passing of our sensory perceptions will do to man's creativity. Can we have artists who do not *see* or poets who are not moved by natural beauty?

I am reminded of René Dubos's wise comment: "Man . . . can adapt to almost anything. That is the real tragedy. . . . As we become adapted we accept worse and worse conditions without realizing that a child born and raised in this environment has no chance of developing his total physical and mental potential."

The next generation of children that grows up on this farm will not have the same advantage as the generations that went before them. They will never know the peacefulness of this unspoiled scene nor look forward to the end of a summer day when they can watch the golden light of reflected sunlight spread across the fields. They will have to learn not to see and not to listen in order to live with the irritation of a constant buzzing noise and the flicker of blue lights along the high-power line—because, gigantic as these towers and conductors may be, it turns out that *they are not large enough*. There will be a continual discharge of electricity into the air and this discharge will be both audible and visible.

Electrochemical Pollution

It may come as a surprise to many people to hear that the lines that carry these enormous voltages are not insulated. In fact, no transmission lines are insulated like the electric cords in our homes. The bare wires pass overhead, in many cases, no more than 35 to 40 feet above roads and farms. The theory is that air is a good insulator, and this is true up to a certain critical voltage. Beyond this point any increase in voltage causes the air to break down as an insulator, and electricity is discharged into the air. The critical point at which this corona discharge occurs depends on a number of factors inherent in the design of the line, such as the diameter and spacing of the conductors. In general, the larger the diameter of the conductors and the wider the spacing, the higher the voltage that can be carried without reaching the critical voltage.

There is, however, a practical limit to the width of spacing that can be achieved on an overhead structure. Conductor size also has natural limitations. An advantage can be gained by using "bundles" of conductors instead of single ones, and this advantage increases with the number of conductors in each bundle. But it is offset by the greater weight, complexity, and expense of the construction. The diameter of the conductors used in the bundles is also a very significant factor in determining the critical voltage of that line. For a given design and spacing the diameter of the conductors should increase as the applied voltage increases if electric losses from conductor to air are to remain constant.

These losses are quite different in nature from the heating losses due to the resistance of the conductor. In this case the electrons are actually passing out into the atmosphere. You can visualize what is happening in a very simplified analogy by imagining the electricity as a stream of electrons racing at enormous speeds along the conductor (they travel mostly on the outside surface of the metal). A conductor with a large diameter gives them a less precipitous path to follow. To the electrons, a large wire looks like a highway with gently sloping shoulders, while a narrow wire looks like a mountain trail with sheer drops on either side. Now imagine that the electrons are being pushed and pressured from behind—the greater the pressure the more likely they are to stumble and fall off into the abyss. Voltage is electrical pressure, so when you increase the voltage you have to provide a broader highway or you will get an increasing number of fatalities along the line. Any roughness of the surface of the conductor also increases the likelihood of losing some of the electrons, just as rocks and ruts along the mountain trail increase the likelihood of a disastrous fall as you speed along it.

No man-made conductor can be smooth enough to look perfectly uniform to objects as small as electrons. There are inevitable variations along the surface even before the line is installed. After it has been in use for a while it becomes weathered, corroded here and there, and encrusted with particles of soot; and these irregularities are more significant. Rain and snow have an even larger effect on the geometry of the line. Raindrops collecting under the line look like upside-down mountain ranges to the electrons traveling along the conductor; they increase the probability of the electrons leaving the surface. These variable factors cannot be controlled, but a line that is designed to operate with a considerable safety margin will not often produce corona discharge.

As transmission voltages have steadily risen, it has become increasingly difficult to use conductors of large enough diameter to provide an adequate safety margin. Engineers are faced with the mechanical problems of supporting larger and larger structures. The wires become stiffer to coil and heavier to string. One might assume that these facts would eventually impose a natural limit on the extent to which voltage increase can be carried.

There is, however, another way of looking at these same facts, which offers a striking example of what one might call the "technological mentality" at work. If the situation is studied only as a cost analysis problem, it turns out that smaller conductors and towers are sufficiently cheaper than those theoretically required for the job so that the companies can lose quite a lot of electricity and still come out ahead. As they go to higher and higher voltages *without increasing the cost of the line* in direct proportion, they decrease the unit cost of electricity. Therefore, they can afford to lose more and more of it, and so on—in an ever-increasing spiral.

The rationale for this type of engineering goes back more than sixty years to Lord Kelvin. His "law" stated that the economical size for a conductor is that for which the additional annual charges on the investment exactly equal the additional annual cost of the energy lost. Lord Kelvin lived in an era when prodigality with natural resources was the rule in industry and the discharge of waste products into the environment was not recognized as constituting a danger for society. The power companies are still using this wasteful and oversimplified criterion in choosing their designs.

The principle of economy through deliberate waste is built into our whole industrial system. For instance, it is cheaper to build automobile engines that have a poor thermal efficiency than to build engines that would achieve a more complete use of the fuel. The cars can be sold cheaper; so more cars are sold. Mass production reduces the unit cost, and so on—another spiral dependent on the principle of economy through waste. The examples are endless. It is cheaper to produce power by burning high-sulphur coal without attempting to utilize the chemical products of the combustion. It is cheaper to throw away beer cans than to collect them and recycle the metal. It is cheaper to allow old automobiles to rust in graveyards around the countryside than to transport them to a factory and salvage the materials.

However, there is a hidden price for these economies. The price is paid by all of us in the form of a steadily increasing load of waste products poured into our atmosphere, our lakes and rivers—the exhaust fumes from automobiles, the smoke from factories, the piles of trash.

What can we do to prevent these wasteful spirals from becoming established? It is certain that the earlier they are identified the better chance there is of controlling or redesigning the processes to yield more acceptable solutions. As an industry becomes more deeply committed financially to one mode of operation, its resistance to change rises exponentially. So whenever a new technological breakthrough is announced—a cheaper, more "efficient" production method—we should look to see what is being thrown away in the process. Is the waste product derived from one of our non-renewable natural resources? Is it an item that may become scarce in the foreseeable future? Is it something that will degrade our environment? When these factors are considered many of the cheap ways of doing things turn out to be more expensive than advertised.

At first the amount of the waste product produced may be small and the effect not readily apparent. This is true of the electric waste on our transmission lines. The majority of the lines we see today are designed to operate at about 50 or 60 per cent of critical voltage. If they are properly maintained they rarely produce corona discharge except under extreme weather conditions. However, the higher voltage lines are being designed to operate closer and closer to the critical voltage and produce corona discharge for a larger percentage of the time. Because of the conductor size and spacing chosen for the first

765-kv lines constructed by the American Electric Power Company, they operate at about 83 per cent of critical voltage. With this high an operating-to-critical-voltage gradient the very smallest discontinuities, such as tiny scratches or grains of pollen on the line, cause corona discharge. These factors are always present, even in fair weather. Therefore, there is a constant crackling and humming sound and, at night, blue glow along the line. Rain or snow storms result in light and sound effects resembling heat lightning over Niagara Falls as electricity pours into the atmosphere.

Unless action is taken to reverse this alarming trend in high-voltage transmission design, the discharge of electricity into our atmosphere will occur in ever-escalating amounts as more 765-kv lines mushroom over the countryside. And even these are just the beginning. The electric utilities are already planning transmission lines that will carry 1,500,000 volts.

It is important to recognize, however, that it is not the transmission at higher voltages *per se* that causes the unnecessary waste and pollution. Transmission at high voltages offers an efficient way of moving electricity; and transmission lines can be built that carry these voltages without such heavy losses. The fault lies with the principle on which these lines are designed. This principle dictates the choice of the cheapest possible construction in order to transport electricity with maximum economy, trading the cost of the lost electricity for the cost of the better construction. This economy is achieved at the expense of our environment.

Electricity is derived from our non-renewable natural resources, which are expected to be in short supply in the next few years. When energy enters the atmosphere in the form of high-voltage corona discharge, it initiates electrochemical processes known to be harmful to all living things, and at the present state of our knowledge there is no way of removing these chemicals from the atmosphere at any price.

In corona discharge, high-energy electrons leave the surface of the conductor, strike the molecules of the air, and cause them to split into molecular fragments. Air is normally composed of molecules of nitrogen, oxygen, and other gases in a relatively quiet, unreactive state. Each gas molecule is a well-balanced entity and therefore has an intrinsic stability. Now when such a molecule is struck by a fast electron coming off a high-voltage conductor, several different things may happen. The impacting electron may knock off one of the electrons orbiting the nuclei of the atoms. This produces a free electron and a molecular fragment called an ion. Another possibility is that the impacting electron can impart additional energy to one of the orbiting electrons, which creates an excited molecule. But an excited molecule is an unstable arrangement and may break down into single atoms. These single atomic configurations are called free radicals. They are more stable than excited molecules but not as stable as the original gas molecule.

In general, all the products of corona discharge are more reactive and have higher energy content than the starting products. The extra energy, of course, comes from the electrons that pour off the high-tension lines. These bundles of energy are passed back and forth like hot potatoes between the various molecular forms as the effect diffuses into the surrounding air.

These physical changes may involve any of the various types of molecules present in the atmosphere, which is a mixture of many different constituents. Oxygen and nitrogen are the principal ones, but there are traces of other elements as well—helium, neon, argon, krypton, and xenon. Several compounds, such as carbon dioxide and water vapor, are always present. Various impurities like sulphur dioxide and carbon monoxide are present in varying amounts, particularly around industrial areas.

When we remember that all the processes mentioned above can occur with most of these molecules and that each of the fragments can enter into combination with a large number of the other molecular fragments, we can appreciate the great complexity of corona chemistry. The air surrounding the discharge becomes a veritable seething cauldron of electrical and chemical activity. It is filled with excited molecules, free electrons, free radicals, heat, and light, which creates the characteristic blue glow. The effect spreads rapidly into the surrounding air as the free electrons go on to impact other molecules and as the molecular fragments seek out other fragments with which to recombine. There are many processes involved, and a few of the chemicals formed are considered to be particularly damaging to living things.

One of the most important reactions involves the element oxygen, which comprises about 21 per cent of the atmosphere. The normal oxygen molecule contains two atoms; when it is struck by a high-energy electron, excited molecules, free radicals, and ions are formed. These combine with normal oxygen molecules to form a special type of oxygen molecule containing three atoms. This molecule, known as ozone, is much more reactive than oxygen. Ozone can even be explosive under certain conditions.

It has long been known that the action of the ultraviolet component of sunlight on the earth's atmosphere produces some ozone. This reaction occurs primarily at the uppermost level of the atmosphere where the energetic ultraviolet rays first encounter the oxygen molecules. The layer of ozone formed by this process absorbs the ultraviolet radiation and thus acts as a screen, filtering out the highest-energy portion of the solar radiation. This makes it possible for life to exist on earth; if the earth were not protected by the layer of ozone, the ultraviolet portion of the solar radiation would destroy the fragile organic molecules that comprise all living things. It is also fortuitous that the ozone layer occurs very high above the earth's surface so that we are not directly exposed to high concentrations of this extremely reactive chemical.

Eventually a small proportion of the ozone molecules formed at high altitudes find their way down to the lower atmosphere. In addition there is a little ozone formed all the time at our level. Simultaneously, some of the ozone is gradually converted back into oxygen or is used up in other chemical reactions. The net result of all these processes is that our atmosphere has a very minute but measurable component of ozone, about .01 to .03 parts per million. Normally the proportion tends to be higher in rural areas than in cities, perhaps because of the more direct action of sunlight. Recently, however, it has been increasing in cities, and the discovery has been made that certain polluting chemicals help to start the process of ozone formation.

Ozone has a distinctive odor. In fact, the name is derived from the Greek word *ozein,* to smell. Its fresh, pungent odor similar to chlorine is pleasant to some people. This fact plus the association with rural environments led to the general belief that ozone was a beneficial and revitalizing element in fresh air. People went to the country to breathe the ozone. But in the last decade more careful study has led to a complete reversal of this opinion. Ozone is now recognized to be one of the most toxic elements in our atmosphere and the proportion of it is increasing due to the various ways in which man is tampering with the environment.

The man responsible for identifying the role of ozone in air pollution was the Dutch scientist A. J. Haagen-Smit, who had come to the United States about twenty years earlier and had been pursuing his research on plant hormones at the California Institute of Technology. Then one day as he was hurrying to his laboratory through a typical smoggy Pasadena morning his attention was suddenly focused on a peculiar odor that pervaded the atmosphere. Many people, of course, were complaining about the smelly gray blanket of smog that occasionally settled over the Los Angeles basin. But Dr. Haagen-Smit's nose was unusually well-educated because odors had been occupying his attention for the past few years. He had been investigating the chemicals that control odors in plants and was concerned with such questions as why onions smell different from pineapples. But the odor that caught his attention that particular morning was not a sweet fragrance like pineapple; it was sharp and acrid like chlorine.

Before the early 1950's the standard explanation of the smog in the Los Angeles basin was that it derived principally from factory smokestacks. But why should factory emissions produce a chlorine-like smell?

To the curious mind of a scientist a question like this must be answered; and furthermore, Dr. Haagen-Smit had the equipment to make the chemical analysis of the air in his laboratory. A condensate of an air sample proved to contain water with a number of strong-smelling compounds—acids, aldehydes, and peroxides. None of these compounds were thought at the time to be products of factory emission. This discovery diverted Dr. Haagen-Smit from pineapples into a lengthy investigation of the chemistry of air pollution.

One of the results of his research was the discovery that ozone is present in surprisingly large amounts during attacks of smog conditions in the Los Angeles area. It sometimes reaches concentrations about twenty times the normal percentage in the atmosphere. The reason for this large build-up of ozone appears to be a chain of chemical reactions involving the waste products of automobile exhaust (and to a lesser extent, factory and power plant emissions) and the action of sunlight. At our atmospheric level the sun's rays are not energetic enough to create ozone directly from oxygen but they are energetic enough to break down one of the oxides of nitrogen, starting a cycle of chemical reactions in which ozone is produced as well as free radicals of oxygen and hydrogen compounds.

The products of this photochemical reaction are very similar to the products of corona discharge. This fact is not really surprising since both those phenomena involve the addition of extra energy to molecules in the atmosphere. In corona discharge the energy comes from the impact of high-speed electrons; in photochemical reactions it comes from the ultraviolet portion of the sunlight. Ozone, nitrogen dioxide, and various free radicals belong to a family of chemicals known as *photochemical oxidants, photo* because they are created by the action of light and *oxidant* because of the type of chemical reaction they initiate. Many of the most violent and destructive reactions that we encounter everyday are examples of oxidation: the burning of wood, the bleaching of colors, the rusting of metals, the explosion of gunpowder. These reactions all occur with the release of energy.

In air pollution chemistry, photochemical oxidants are usually measured together and reported as "total oxidant." Since Dr. Haagen-Smit's pioneer work a great deal of study has been devoted to this group of chemicals. Ozone is found to comprise about 90 per cent of the total oxidant in air pollution, and has therefore received the most attention. However, several of the lesser members of the family are especially interesting and may prove to have biological significance far out of proportion to their prevalence.

One comparatively rare form of oxygen has recently been recognized as a very lethal factor in air pollution. This is an excited state of the oxygen molecule known as *singlet oxygen.* Like other excited molecules, it is highly reactive and intrinsically unstable so that, on the average, it exists for only a tenth of a second. This lifetime may seem too short to be significant; but it is long enough for the molecule to enter into biologically destructive reactions. Singlet oxygen is produced directly from corona discharge acting on normal oxygen and also by the reaction of ozone with either normal oxygen or organic molecules. Much further research is needed to understand the role of singlet oxygen but at present there are strong indications that it plays an important role in smog formation and that, furthermore, there is a connection between singlet oxygen and cancer. Dr. Ahsan Ullah Khan of Florida State University has proposed a mechanism by which singlet oxygen can attack DNA or an enzyme to produce malignant tumors. According to this theory, ozone breathed in from polluted atmosphere can react with certain organic compounds to produce singlet oxygen inside the body. These possibilities appeared real enough to bring 135 scientists from the United States and foreign countries to New York in October 1969 for an international conference on the role of singlet oxygen in the environment. Work is now in progress in many laboratories on this rare molecule and we will probably hear more about it in the near future.

Nitrogen, the most common element in air (78 per cent by volume), is one of the most neutral and benign substances in its normal state. It is colorless, odorless, and nontoxic, serving as a diluent for the much more reactive element oxygen. However, under electric discharge or in high temperature combustion (as in automobile engines and steam boilers used to generate electricity), nitrogen combines with oxygen in various ways, the principal product being nitric oxide. In itself nitric oxide is relatively harmless, but it reacts with ozone to produce nitrogen dioxide, the acrid, whiskey-colored compound of smog. Nitrogen dioxide has been found to cause lung tissue damage in laboratory animals and to cause increased susceptibility to respiratory infection.

The presence of this oxide starts the cycle that Dr. Haagen-Smit discovered. Nitrogen dioxide is broken down by sunlight (or electric discharge) to atomic oxygen and nitric oxide. Atomic oxygen combines with oxygen to make ozone. The nitric oxide is converted again into nitrogen dioxide and thus a cycle is set up, resulting in considerable levels of these pollutants remaining in the atmosphere.

Nitric oxide can also react with water to create nitrous acid, which is known to cause genetic mutations in plants and lower organisms. There is considerable concern that nitrous acid in the human system may produce cancer. This concern led to the banning of the use of sodium nitrite in foods, because it has been shown that sodium nitrite is converted to nitrous acid in the stomach.

Finally, there is a class of organic compounds containing nitrogen (as well as carbon, hydrogen, and oxygen) suspected of playing an intermediary role in the formation of toxic smog. This group of compounds is known as PAN (short for peroxyacyl nitrates). The formation of PAN in the atmosphere is not very well understood but it is thought to involve free radicals of oxygen, nitrogen oxides, and organic molecules. PAN is destructive to vegetation at extremely low concentrations.

All of these highly reactive and toxic chemicals can be produced in the breakdown of air by electric discharge. Yet electric companies build thousands of miles of lines that create corona discharge without any attempt to evaluate the effect of these chemical reactions on the atmosphere along the rights-of-way where many people will spend a large part of their lives.

Noise Pollution

Most of the dangerous effects of electric discharge take place without any noise or visible signs to warn of their presence. However, certain phenomena associated with corona discharge are readily observable. When the air becomes a partial conductor, electromagnetic fields are set up which interfere with radio and television signals. Ever since high-power transmission lines have been used, complaints about interference have been received by utility companies and, therefore, they do pay some attention to these problems in designing their lines. But in order to reduce the radio interference, heavier and more expensive construction is required. It is to the financial advantage of the electric companies to build the lightest line that will be tolerated by the people living nearby. The engineers' estimate of the maximum interference that the public will accept is known as their "criteria of acceptability."

In discussing radio reception the important relationship to keep in mind is the ratio between the signal and the noise levels. The larger this ratio, the better the radio reception. If the strength of the background noise is close to the strength of the signal then, no matter how high you turn the volume, you will not be able to distinguish the program from the noise.

Because of their proximity to radio stations, urban and semi-urban areas have signal strengths much greater than do rural areas. There is always a certain amount of background noise or static. Local atmospheric conditions—storms, electrical disturbances such as corona discharge—cause the background noise to increase. The standard classifications of radio reception are described as: Class A, entirely satisfactory; Class B, very good with background unobtrusive; Class C, fairly satisfactory with background plainly evident; Class D, background very evident but speech easily understood.

According to the published estimates of the American Electric Power engineers, the signal-to-noise ratio at the edge of the right-of-way of their 765-kv lines in rural areas was expected to be very much poorer than the ratio of even Class D radio reception. And in foul weather the noise level was expected to be greater than the signal.

The engineers attempted to justify their "criteria of acceptability" on the following grounds: they hoped to maintain an average distance from line to dwelling of about 200 feet, at which distance the fair-weather radio reception in semi-urban communities would be "fairly satisfactory with background plainly evident." Foul weather reception would be Class D or worse.

There are three important reasons why these criteria are not acceptable:

1. The majority of the people affected do not live in semi-urban communities. The lines have been sited wherever possible through rural and remote rural areas in order to avoid the expense of rights-of-way through suburban property or the expense of underground construction.

2. The line can pass closer than 200 feet to many homes. The electric utility companies buy a right-of-way 200 feet wide (100 feet on each side of the center conductor). When they speak of the average distance from dwelling to line, they include the fortunate homes that happen to be quite far removed from the line. Obviously, if the *average* distance is 200 feet many homes are closer than 200 and some may be right at the edge of the right-of-way. If so, the owner has no legal protection even though his radio reception is entirely destroyed. Furthermore, the land within a 100-foot strip on either side of the right-of-way will never be suitable as future building sites.

3. Bad weather prevails nearly a fifth of the time. In the country, people are apt to be out of doors in fair weather. It is during foul weather that they turn to their homes and depend upon such entertainment as radio and TV.

Just in case these "criteria of acceptability" prove too generous, however, there is a way in which the standards can be even further reduced. The power companies have discovered that lines designed for one voltage can be run at a slightly higher voltage, thereby achieving an absolute peaking out of economy. For instance, Ohio Valley Electric Company designed lines to carry power from Kyger Creek to the Atomic Energy Commission's gaseous diffusion plant near Piketon. These lines were designed to carry 330,000 volts but they have actually been operated at 345,000 volts. "The transmission system proved its design specifications and then some," they proudly report. "Operation at 345,000 rather than 330,000 was not only feasible, but even more economical." This increase in voltage results in a very large increase in radio interference and corona discharge.

I have driven under these lines from Kyger Creek. At a distance of several hundred feet, the music on the car radio was drowned out by noise like a million angry bees. The American Electric Power engineers are frank in saying that they expect to increase the voltage on the 765-kv lines. At first they will operate slightly below this level until all the problems have been ironed out (such as complaints of the property owners along the line). Then the voltage will be stepped up until the line is operating at 775 or 800 kv, thus further overloading an already under-designed line.

Television reception is also affected by the electromagnetic fields occurring around high-tension lines. Interference takes the form of bright streaks or bands that drift slowly across the screen. Certain channels are much more severely affected than others. The frequencies used by Channels 2-6 are especially sensitive. Interference on Channels 5 and 6 has caused many complaints.

American Electric Power Company expects to deal with the complaints that arise by means of a truck set up as a traveling laboratory to test and "correct" reception problems where they arise. In remote rural areas with weak signal strength "corrective measures" will be applied. The measures include the installation of huge antennas on the rooftops and on nearby hills—more steel scarecrows to match the towers in the fields.

The truck, of course, cannot be everywhere at once. During a storm it can answer only a very few of the calls for service. But after a while most people stop complaining. It is the experience of the utilities that "many people will not initiate a complaint but would say that the interference was not acceptable if we asked them." Eventually the almost infinite adaptability of man will assert itself. People will get used to looking at screens filled with streaks and obscured with "snow."

In reality, the main object of the little traveling laboratory is to help evaluate this project which represents one giant experiment—an experiment involving millions of dollars and many thousands of people as guinea pigs. The designers hoped to "obtain more information with regard to TV interference as soon as the first of the 765 kv lines is energized." They also expect to acquire "experience useful to possible reevaluation of radio interference bench marks for future and higher voltage systems. . . . An immediate benefit is the utilization of conductors which are economically closer to the system load transfer requirements." In plain English, those statements mean that they did not understand all the factors involved in TV and radio interference from corona discharge. They expected to learn about these *after* they had built the line. They also intended to find out just how far they could go in exploiting the economy of this type of transmission before they encountered serious resistance from the American people.

It is a cherished belief in this country that we have a higher standard of living than anyplace else in the world. So it is surprising to learn that in some other countries the public interest is more respected in these matters than it is here. Canadian engineers commented that American Electric Power's 765-kv lines would produce more corona discharge and more radio interference than any other lines in the world. The levels of radio interference, they said, would not be tolerated in many parts of their country. "We would not dare to build such lines in Canada; the public outcry would be too great."

With the introduction of the 765-kv lines a new noise problem became a significant factor. These lines produce a continuous humming and crackling sound, which in rain or snow becomes a loud roar. These audible effects are caused by the impact of air molecules, like miniature thunderclaps. Surges of high-speed electrons and charged molecules create little vacuum pockets in the air and other molecules rushing in to fill these vacuums collide, producing the sound.

Under the new 765-kv lines, noise levels of up to 70 decibels have been recorded. At this level it is necessary to shout to be understood in a normal conversation. Ninety decibels is the legal limit for noise levels that may be regularly imposed upon people during working hours. Beyond that level employees must be protected to prevent deterioration of their hearing. Many city dwellers are now protected against irritating noise levels. For instance, in Chicago a city ordinance forbids the installation of appliances that cause noise levels exceeding 55 decibels at the lot line in residential zones.

The first 765-kv lines that were energized produced more audible noise than the designing engineers had anticipated. Since audible noise is directly related to corona discharge, it is reasonable to assume that the amount of corona discharge is also greater than they anticipated. The power companies are receiving many complaints from the people living near these lines. The noise wakes them up at night and even on a fair day is a constant source of irritation. Unlike radio and television interference, this audible noise does not diminish rapidly as one moves away from the line. It changes very slowly with distance and, therefore, more people are exposed to objectionably high levels. The power companies are planning to go to even higher voltages in the very near future. Audible noise for voltages of 1000 kv, 1200 kv, and 1500 kv will present more serious problems. But the utilities do not intend to allow this side effect to interfere with their plans for going to these higher voltages. The most economical design for these lines will be determined by the audible noise levels that the engineers believe people can be forced to accept. They are talking already about considering 60 decibels at the edge of the right-of-way an acceptable noise level.

Second-Class Citizens

The attitude of the power industry is that it is permissible to impose these unpleasant effects on rural citizens living along their right-of-way. What if a few farmers must suffer a little inconvenience of this kind in order that the rest of the people may benefit? But this attitude violates the constitutional principle of equal application of the law. The laws enabling the public utilities to appropriate land for rights-of-way make it possible for them to impose environmental damage on rural property owners that would not be tolerated in more densely populated areas.

In fact, the criteria of acceptability used by the power companies make it apparent that they are not concerned with protecting the rights of rural citizens. Their radio and television interference standards, for instance, consider only the requirements of urban residents. "The present basic philosophy," say the AEP engineers, "places almost all of the weight on selecting a specific conductor diameter that would allow satisfactory radio reception during fair-weather conditions in urban areas." The application of these criteria to line design means that the rural resident, living in areas where signal strengths are lower, will inevitably suffer. At the same time the lines are being routed to avoid urban areas.

These design practices clearly discriminate against the rural citizen and are recognized as being discriminatory by the industry. A professional group of electrical engineers made the following comment in a report on radio noise-level criteria: "The noise levels quoted . . . raise an interesting question concerning the right of a rural listener to expect quality reception equal to (or better or worse than) his urban counterpart. Should additional monies be spent on transmission lines where only a few residences per mile are encountered, even although the broadcast signals are at a low level? At what distance from the right-of-way should one expect to attain quality reception? The acceptability of recently completed extra-high-voltage lines will have to be assessed in the next few years. . . ."

Thus the power of eminent domain makes it possible for the power companies to distribute high levels of visual and audible pollution across the countryside arbitrarily, while the benefits of cheaper power are funneled into the large industrial centers and sold at wholesale rates to large consumers. The degree of protection of an individual's rights depends upon where he lives (a concept that is certainly at odds with the principles on which this country was founded). By these practices the electric utilities are destroying rural values to increase the wealth of the cities. Furthermore the number of disadvantaged citizens is increasing every year. The network of lines already constructed is approximately 1200 miles long and each mile intercepts, on the average, four or five pieces of property. So we could estimate that this network now involves approximately 25,000 people. And this is just the beginning. According to official government projections, approximately 100,000 miles of new transmission lines will be constructed each decade for the balance of this century. By the year 2000 this network will have expanded into a vast steel spiderweb enmeshing our countryside, and the people who live beside these lines will number in the millions.

Remember also that once the investment has been made in these lines, they will cast their shadow far down into the future. In the meantime our civilization is becoming increasingly dependent on electronic modes of communication. Even today in remote country areas they actually constitute the major source of information and cultural exposure. Rural residents depend on television and radio to keep them in touch with the rest of the world, provide recreation in bad weather, and alleviate the loneliness of long winter nights. At these times severe interference will virtually cut them off from this resource. It will also mean that rural residents cannot enjoy many of the new electronic conveniences. Solid-state technology is rapidly enhancing the advantages of small portable sets. But these sets will not be usable in the majority of homes along the EHV (extra-high-voltage) transmission lines. Electronic intercommunication systems, educational and shopping services via TV—these and other devices projected for the near future will be subject to the same degradation of quality in these locations. Yet these conveniences will mean more to country people than to urban residents who already have goods and services at their doorstep.

Imagine that you live in the hill country near Vinton, Ohio, perhaps in the wooded valley of Little Raccoon Creek. The nearest movie theater and public library are seventeen miles away by poor roads; the nearest department store is fifty miles away; the nearest legitimate theater and art gallery are a hundred miles away. Several times during the winter, snowstorms may make it impossible to reach the main roads at all. Imagine how you would feel if on those snowy nights you found you were unable to get anything more than moving streaks across the TV screen.

As time goes on the presence of high-corona transmission lines will cause an increasing deterioration in the quality of rural life and will help to accelerate the exodus to cities, contributing to the increasing density of urban population, which is already becoming intolerable. What we should be doing is the exact opposite–improving the convenience and opportunities of rural existence, encouraging people to spread out where there is room to breathe without sucking in one's neighbor.

Above all, we should be preserving the beauty and quiet of these country settings, where a man can escape from the distracting roar of a man-made environment–where he can enjoy the frozen silence of a winter landscape or listen to the myriad tiny living sounds of a meadow on a hot summer day. Living in these quiet country places, a man can lose himself in a larger organic whole and know the serenity that comes from finding his own place in the endless diurnal cycles and the slow turning of the seasons.

Country people do not talk much about their relationship to nature but they know it is their most precious possession. Officials of the power company, accustomed to thinking that anything is for sale if the price is high enough, are baffled by the attitude they frequently encounter: "I don't care that much about the cash. I don't want a million volts crackling overhead and I don't want your ugly towers. I'd rather save my view."

3

The People Protest

WE THE UNDERSIGNED ARE OPPOSED TO THE INSTALLATION, BY THE AMERICAN ELECTRIC POWER COMPANY OR ITS SUBSIDIARIES, OF A 765,000 VOLT POWER LINE ON OR NEAR OUR PROPERTY. WE SPECIFICALLY PROTEST AGAINST THE INSTALLATION OF SUCH A LINE BECAUSE OF THE HAZARDS TO HEALTH AND THE NUISANCES EXPECTED THEREFROM.

Ira and Mary Wolf
Tessie Brown
Earl Gunther

The petition was passed silently from hand to hand around the room. It was a warm Sunday late in May and the doors of the little schoolhouse were open, letting in a broad ribbon of sunshine and the darting shadow of a hummingbird busy in the wisteria vine that hung heavy with lavender bloom over the front porch. Outside in the schoolyard stood two long trestle tables left over from the strawberry social held there the evening before, and beyond the schoolyard stood the white frame United Brethren Church with its cluster of family burial plots. Many of the graves here were marked with crumbling tombstones dating back to pioneer days:

Eliza, daughter of Abraham and Magdalene Ranck, died August 5, 1810. Aged two years, 2 months and 10 days.

Ephriem Hough . . . died July 5, 1820.

Livonia, wife of Nathaniel Kellenberger . . . 1816.

A few of the oldest stones bore German inscriptions from Pennsylvania Dutch families that had come over the mountains and settled here at the end of the eighteenth century:

> Hier Ruhet Catarina Dresback
> Gebornen den Jahr 1730
> Gestorben den 15 Juni 1805

Many of those same names were going down on the petition inside the schoolhouse.

Irma Kellenberger, the postmistress. Her grandfather and father had been postmaster before her.

Ellersly and Mildred Ranck—Mildred was noted as one of the best cooks in the county. If you wanted to buy one of her famous hickorynut cakes at the fish-fry you had to speak for it days ahead.

And old Jennie Dresback. Jennie, who still wore her sunbonnet even on Sunday. She had given birth to twenty-two children and raised eighteen of them.

Damon and Ora Pontius, who grew such delicious strawberries that in season people came from miles around to buy them. Their flavor was as sweet as tiny wild berries and yet they were so large that only four or five could be packed in a pint basket.

Kiziah Hostler Hough—known as Aunt Kiz to everyone in the village. For the past thirty-five years she had taught Sunday school in the Lutheran church, and she was always in the vanguard of every local cause.

Donald and Alice Mueller. Donald was County Health Commissioner and Alice was head surgical nurse at the county hospital. They commuted fifteen miles every weekday in order to live in the village of Laurel where he had grown up and where his father had operated the general store—known in those days as Mueller's Department Store. Alice and Donald had taken one of the oldest houses in the village and remodeled it. From the outside it looked like a doll's house—tiny and neat. Inside it was surprisingly spacious and well appointed.

In their spare time Alice and Donald were birdwatchers. They belonged to conservation societies and had many out-of-town friends. Because the area around Laurel was particularly scenic and attracted a number of unusual birds, other birdwatchers from the nearby cities came and often they visited the Muellers. On weekends and summer evenings they could be seen tramping through the fields and woods armed with binoculars and cameras.

They knew the pond where the kingfisher lived and the nesting place of the horned lark in the copse back of Aunt Jo's hill. They often visited the corner down on the Old Swamp Road where the cool call of the whippoorwill could be heard every summer evening, filling the valley with its tranquil, melancholy sound.

They knew also that the prairie land to the north of the village was rich in wild life; muskrats, rabbits, field mice, and oppossums were abundant there, particularly in the tall grasses and brush by the deep ditch that ran along Whistler Road. Migrant birds attracted by the food and the light cover often stopped there to rest on their long journeys—golden plover flying from Alaska to southern Argentina—snow buntings and Lapland longspur migrating from the Arctic tundra to Louisiana and the Carolinas.

To these birdwatchers the prairie land along Whistler Road was a uniquely lovely place. They had written a description of it for the Audubon Society: "Some of us," they said, "have known Whistler Road for many years, others not so long but all of us have learned to appreciate its stillness of a winterday, the effortless flight of the great Hawk, the covey call of the Bob-White as the day ends and the Short-Eared Owl rises up from the dense grasses and on silent wings hunts for food, as he has done for ages past."

The news that a high-power line would cut a broad swath across the prairie and along Whistler Road had shocked the Muellers into action. Even though they had no property directly involved in the right-of-way, they had been among the first to raise their voices against it. Alice had contacted most of the local people and had called the meeting at the schoolhouse. She had written to the Audubon Society and several conservation societies. Some of these groups had sent representatives to the meeting.

One of the main objects of getting together, Alice explained, was to exchange information. The agents of the power company had been very careful to dispense as little information as possible, thereby making it difficult for neighbors to cooperate in opposing the line. Each landowner had been shown only the path of the right-of-way through his own property.

From the Department of the Interior Alice had obtained detailed Geological Survey maps, and at the meeting each farmer drew in the path of the line across his property. As the route took shape across the geological map, one thing became immediately apparent. The line was routed through a portion of Tar Hollow, the State Forest. A mutter of resentment passed around the room.

"They don't have no right to use park land."

"Our taxes pay for that park."

"We shouldn't stand for such a thing happening to Tar Hollow!" exclaimed Aunt Kiz. "Why don't we all lie down in a row in front of the bulldozers like the Israelites in front of the Roman legions!" The suggestion met with general enthusiasm.

"I heard of an old woman down in Kentucky who stopped the strip-miners that way," Alice said. "Still, I'm not sure these people wouldn't run right over us."

Most of the meeting was spent discussing the dangers of the proposed line. Everyone had something to contribute. "My brother-in-law has a line over his farm," volunteered one of the landowners. "Not so high powered, neither, and he has a real hard time getting his cattle to go under that line. He says the hair on their backs stands straight up on end."

"I have a friend who is on the Public Utility Commission," said another farmer, "and I asked him—'Is it true, all this stuff about a line like that being dangerous?' 'You don't know the half of it,' he told me. 'You can get a shock bad enough to knock you down just getting on and off your tractor under a line like that!' "

"They say all the fences have to be grounded and the barns and the farm machinery—how about the farmers? Are they going to ground us, too?"

The current of anger swirled back and forth as everyone spoke his mind. But the meeting finally broke up on a hopeful note. "I don't believe the government will allow it," said Jesse Snyder. "Not if they know it might do as much harm as all that."

"Sure," rejoined his brother. "It's just a matter of bringing this to the attention of the right authorities. We ought to write our senators and that federal power outfit down in Washington. They'll do something about it. If it's harmful the government won't allow it."

Alice Mueller volunteered to write the letters. And so as the weeks passed a steady stream of correspondence began going out to senators and representatives and government agencies from the little house in Laurel, Ohio.

4

"The Government Will Protect Us"

Weeks of correspondence and consultation with lawyers and government agencies revealed the surprising and shocking fact that in Ohio (as well as in many other states) there was no public body that had the power to pass on the design of electric installations or the responsibility for protecting the health and safety of the citizen in these matters. Furthermore, there is no federal body with jurisdiction over most of these electric installations. The Federal Power Commission has authority only over facilities related to hydroelectric projects. The Army Corps of Engineers has jurisdiction only where lines cross navigable rivers. Even the Environmental Protection Agency has the right to require review of only those projects that are licensed or financed by a federal agency. There is no mechanism by which a citizen or group of citizens can seek protection from installations they consider to be dangerous without incurring large legal expenses. In designing the equipment, choosing the route, distributing the "benefits," the only party that has any rights is the power industry. How did such an extraordinary vacuum of legal safeguards for the protection of the public come about?

The Right of Eminent Domain

The right of eminent domain grants to electric companies the right to appropriate private property (or an interest therein) as is deemed necessary for the generation and transmission of electric power. The declaration of reasonable necessity by the company is accepted as prima facie evidence and the law specifies that, except in very unusual situations, the issue must be resolved in favor of the utility. If any objections are raised the burden of proof is on the property owner. In appropriation proceedings the only issue supported by precedent is the valuation of the property condemned for right-of-way.

A legal commentary contains the following statement: "A broad discretion is necessarily vested in those to whom the power of eminent domain is delegated, in determining what property is necessary for the public purpose, with respect to the particular route, line, or location of the proposed work or improvement; and the general rule is that the courts will not disturb their action in the absence of fraud, bad faith, or gross abuse of discretion."

These extraordinary rights were given to the electric companies in order to make it possible for them to perform a service to the community, to bring electric power most efficiently to the greatest number of people. At the time the rights were granted the electric utilities were young companies. They had an enormous job to do with limited resources and a still undeveloped technology. A maximum of legal protection against obstruction by private citizens was reasonable to speed the distribution of electricity throughout the country. At that time the transmission lines were relatively innocuous, small in size and carrying low voltage. The only hazard that caused concern was possible contact with a live wire, perhaps a wire downed by wind or ice storm. The electric companies were as anxious as the public to prevent this type of accident because it resulted in a complete loss of power along their line and was expensive to repair. Public interest and private enterprise at this stage were working in the same direction. Electrical engineers built better and stronger supports. Accidents were rare; the public was lulled into a sense of security.

As the years went by, however, the economic success and fantastically rapid growth of the electric industry compounded the problem by adding the power of a vast financial empire to the power of eminent domain. The electric industry now possesses the greatest capital wealth owned by any industry in the United States. Today it is difficult to find a law firm or a board of directors of a large company that is not in some manner involved with the power industry. Similarly, many of the men who make decisions at state and federal levels hold office or own shares in the utility companies. This concentration of power has created an unprecedented situation. In this industry we have a federally protected monopoly granted practically unlimited power to force its will on the American people and yet it is a business organized to make money. In many vital decisions it does not put the public good above its own profit and, in a sense, we cannot reasonably expect it to do so.

In many countries of the world public utilities are government operated. They have the same unlimited powers that our utilities have but their aim is to serve the people. If electricity cost a little more to produce and transmit safely with minimum impact on the environment, a government-operated utility would be in a position to make that choice. A utility operated for profit can always be expected to opt for maximum economy unless compelled to do otherwise.

The chief objection to government ownership of public utilities is that efficiency of operation might be sacrificed if the profit motive were removed. But economy and efficiency are the very motivations that have led directly to the problems we are now encountering. An important decision facing the American public today is how to control the public utilities in order to obtain proper consideration of public safety and minimize degradation of the environment. Should we go to complete governmental ownership of the utilities, or can protective mechanisms be set up which will permit private ownership to continue and at the same time ensure responsiveness to values affecting the quality of life in this country?

Unfortunately, the electric power lobby is fighting hard to prevent any restriction of the extraordinary powers and prerogatives the industry now possesses. As *Business Week* magazine put it: "The biggest single fear private utility executives harbor is that the federal government will increase intervention into what they regard as their domain." The intransigence of the power lobby may lead to more sweeping controls than those they presently fear as the people awaken to the dangers of the special privileges now vested in these companies.

Public Utility Commissions

Over the years a number of government agencies have been set up and legislation passed to protect certain aspects of the public interest against discriminatory or dangerous practices by the electric utilities. In forty-seven of the fifty states regulatory commissions have been created which have some degree of jurisdiction over electric companies. (Minnesota, Texas, and South Dakota are the three without such regulatory agencies for electric utilities.) These commissions are responsible for authorizing rate changes and enforcing safety codes adopted by the state in question. In some states the public utility commissions require certification before a power plant or a transmission line may be constructed. However, thirteen state commissions are without any licensing authority at all, and several commissions have this authority only under special circumstances.

Of the twenty-nine states that require formal authorization before power plants and transmission lines can be built, about half require that public hearings be held before a license is granted. In the remaining states hearings may be held if requested by an intervenor or at the discretion of the commission. Even the holding of a public hearing, however, does not guarantee a fair consideration of objections to the proposed installation. The public utility commissions are charged with promoting the safety, health, and convenience of the public. But because the commissions are composed almost entirely of industry-oriented people, they interpret this to mean primarily the provision of adequate electric power at the lowest possible rates.

Public notice of a hearing is usually made only thirty days before the hearing is held. So while the utility company has had years to prepare its case for the hearing, intervenors have only thirty days to prepare their case, to assemble expert witnesses, and to obtain the affidavits which are usually necessary to present evidence at the hearing. As a practical matter this short notice and the legal expenses involved prevent any effective opposition to the power company's plans. The licensing becomes simply a rubber-stamp operation; and after a license has been granted by a state commission, the private citizen has no right to question the design, the location, or the safety of the installation.

Furthermore, a right-of-way is granted in perpetuity and the utility is given a completely free hand in making any repairs, alterations, or additions to the transmission system once it is installed. For this reason, a hearing on the safety of a proposed facility has, at best, only the most transitory value. The specifications can immediately be altered at the discretion of the utility.

National Electric Safety Code

In 1915 a series of recommendations for the design of electric equipment was drawn up by the National Bureau of Standards. These recommendations are known as the National Electric Safety Code. They provide safeguards against shoddy manufacture of electrical equipment, inadequate household wiring, and similar problems. Revised editions also set standards for transmission line design.

However, only certain portions of this code have been adopted as law by the various states. In Ohio, for instance, the NESC standards for transmission lines are binding only in localities where the lines pass near railroads or public communications systems. Furthermore, the code was drawn up before the days of really high-voltage technology. The safety factors considered in the NESC standards for transmission lines are almost entirely concerned with avoiding actual physical contact with the conducting cables. To use these standards for extra-high-voltage equipment is like applying horse-and-buggy traffic laws to a modern superhighway.

With the advent of extra-high-voltage transmission, a number of potential hazards that had never merited serious concern have become significant. Between the transmission lines and the ground there is a strong electric field. If a metallic object such as an automobile or a piece of farm machinery comes into this field, it takes on a voltage that is some fraction of the total voltage between the conductors and the earth. And if the object is insulated from the ground (by rubber tires, for example) a charge builds up. Then when the object is suddenly grounded by a person touching it, a current flows through the person to ground, the amount of current depending in a complex way on the electrical characteristics of the object, its size and shape, and its distance from the transmission line. The larger the area of the object and the closer it is to the line, the more current will flow when it is grounded. The magnitude of this current determines the seriousness of the electric shock sustained.

If the current is large enough, the person is not able to release his hold on the conducting object and current continues to flow through his body. The threshold of danger is determined by the current a person can tolerate when holding the charged object and still be able to let go of it using the muscles directly stimulated by the current. The safe "let-go" threshold includes a safety margin and is given as 9 milliamps for men, 6 milliamps for women, and 4.5 milliamps for children.

Dr. Charles Dalziel, who has made very extensive studies of these effects, says: "If long continued, currents in excess of one's let-go current, passing through the chest, may produce collapse, unconsciousness, asphyxia, and death. Ventricular fibrillation is probably the most common cause of death in electric shock cases, and may be produced by moderately small currents that cause derangement of coordination within the heart rather than physical damage to that organ. When fibrillation takes place, the rhythmic pumping action of the heart ceases and death rapidly follows."

Calculations made by electrical engineers and published in professional journals show that it is theoretically possible for a person coming into contact with a long metal object such as a pipeline, a fence, or a gutter under a 765-kv transmission line at the minimum heights now in common use to experience charging currents exceeding a man's safe let-go threshold. For a child, of course, the hazard is greater. Currents over 0.5 milliamps—far below the let-go threshold—are also considered dangerous because they may cause involuntary movement and trigger a serious accident.

A working group of electrical engineers made a study in 1971 of electrostatic effects and tested shock from various-sized vehicles parked under high-voltage transmission lines. Their report states that lines carrying voltages *under 500 kv* do not produce shocks that cause major physiological damage. They recommend that parking of vehicles on rights-of-way for lines of 230 kv or less could be safely allowed under controlled conditions, but that the parking of vehicles under lines carrying higher voltages "should be reviewed on an individual basis." Yet people who live next to the rights-of-way for 500- and 765-kv lines are not cautioned about such dangers. Children are allowed to play under the transmission lines. Farmers are told that there is no hazard in parking large pieces of farm machinery under the line; and to lull their instinctive fear of these effects they are told that any shocks received would be "similar to touching a doorknob on a cold day."

It is a strange and awesome sensation to walk under a fully energized 765-kv line and feel oneself coming into the strong electric field. The hair on one's arms stands up. There is a feeling of stimulation and tension in the air like the atmosphere just before an electric storm. The long-term effects of living most of one's life in such an unnatural environment have not been scientifically evaluated, but it is reasonable to suppose that the biological effects may be profound. Each living cell has electromagnetic fields associated with it and biologists have only the most fragmentary information on the influence of external electric fields on the operation of the cellular organization. Pasteur believed that an understanding of the significance of strongly directional external fields might provide an important insight into the nature of living things. Now, a century later, these questions are still unanswered, and many people are being subjected to these environments before their influence is entirely understood.

In answer to questions concerning the safety of the NESC standards as they applied to extra-high-voltage transmission, the following statement was obtained from the National Bureau of Standards:

> The clearances in NBS Handbook 81 were promulgated long before the development of 765kv transmission systems. A subcommittee to reevaluate clearance provisions is now being organized. However, it will be some time before any revisions are made to the code in this area. Generally, changes to the National Electrical Safety Code are not made until after there has been a fair amount of experience with a new development and, insofar as the code is concerned, transmission at 765 is still probably in the developmental or experimental stage.

This letter was written in May 1970, after 200 miles of this type of line had been operating for a year. Three other lines of the same design were under construction. Yet the code that is supposed to protect the citizen against danger from such installations cannot be changed until "after there has been a fair amount of experience with a new development." Apparently, damage must be demonstrated *before* protective laws are passed. In the meantime thousands of unsuspecting citizens are allowed to serve as guinea pigs for each technological experiment.

The National Electric Safety Code specifies minimum heights to be maintained over several different classifications of land use: railroads, public streets, driveways, and "ways accessible to pedestrians only."

There is practically no place in the United States today that is accessible to pedestrians only, but that is still one of the major classifications listed in the code. Farmland, on the other hand, is not specifically listed. Fifty years ago these classifications were reasonable because farms were usually accessible only to pedestrians and animals. But now some utility companies take advantage of the ambiguity of this antiquated code by using the pedestrian-only category to apply to farmland even though modern farms are highly mechanized. Large pieces of farm machinery—tractors, grain elevators, combines, corn pickers—may stand 15 feet high and often pass directly under the line in the course of farming activities. Because the build-up of static charge on vehicles depends on their size and their distance from the conductors, transmission lines over farmland should maintain at least the same minimum clearance as they do over roads.

The American Electric Power Company engineers designed their lines to come within 40 feet of the ground over farmland, 45 feet over roads, and 50 feet over railroads. From these figures it is apparent that the minimum height over farmland is planned to satisfy the NESC standard for ways accessible to pedestrians only. Other electrical engineers, commenting on this design, questioned the propriety of this interpretation of the code: "We are . . . interested to note the 'ground clearance' which seems to be that specified as 'spaces or ways accessible to pedestrians only.' It has been our practice in recent years to pass over this type of clearance since terrain falling strictly into this classification is not usually traversed by transmission lines. . . ."

Attention was also drawn to the fact that no safety margin had been added for the clearance over railroads. Safety margins of several extra feet are normally added to minimum standards to allow for slight errors in survey or construction. In answering this comment the American Electric Power engineers say frankly, "This margin was not included for railroad crossings since the clearance value determined from the code seems overgenerous."

The omission of this margin means that because of construction or surveying errors the minimum distance over railroads will sometimes be less than the minimum required by law. In these cases the lines are subject to state regulations because they cross railroad tracks. Such nonconforming lines have been built and are now operating. More are on the drawing board. Why are lines permitted that do not conform with state law?

Unfortunately, the law is not enforced. Since there is no government agency to review the design before construction, failure to comply with the code is not anticipated. After the line is built no one checks it. The power company knows from experience that no authoritative measurement will be made of the exact distance between the railroad tracks and their lines carrying nearly a million volts. They take advantage of this legislative vacuum.

In Canada, where the electric utilities are government operated, the Canadian Electric Code designates a special classification for "farmland likely to be traveled by vehicles." Their highest-voltage line, which carries 735 kv, has a minimum clearance of 45 feet over farmland.

The power company engineers attempt to solve the problem of electrostatic charge by grounding all the conducting surfaces—fences, corn cribs, metal roofs, even clothes lines—not only in the right-of-way, but also within one hundred feet of its edge. Grounding, however, protects only from charge induced on stationary objects. It does not protect the man who attempts to fill a gas tank under the line. The induced charge under these circumstances could cause a spark and a fatal explosion. Grounding does not protect the boy with a model airplane, or a child with a kite. These hazards will increase in magnitude as the power industry goes to even higher voltages—1500 kv is considered "a certainty" and 2000 kv is "within reason."

Air Quality Standards

Increasing public concern about air pollution has resulted in a number of new regulations at federal, state, and local levels. Under the terms of the Clean Air Act of 1970, each state has the responsibility of drawing up its own set of air quality standards, which must meet minimum National Air Pollution Control Administration Standards. The implementation and enforcement of the standards is at the state level.

In December 1970, Ohio adopted Ambient Air Quality Standards for photochemical oxidants (as well as for carbon monoxide and hydrocarbons). Demonstration of compliance with these standards requires an accurate knowledge of the amounts of oxidant present on an hourly basis over a year's time. However, there is no adequate theoretical basis for computing the concentrations of oxidants that may be expected to build up near high-voltage lines under varying electrical and weather conditions. The problem is a complex one involving wind speed, temperature, humidity, and the presence of other pollutants in the atmosphere, as well as the complicated chemistry of electric discharge itself.

One might suppose that the electric companies, planning to go to lines with higher and higher corona losses, would have devoted some time and thought to these effects, but they have not. The first four 765-kv transmission lines were built without any consideration of the air pollutants they might create. Now, in order to demonstrate that the air quality standards are satisfied and that new lines presently being constructed will also satisfy the laws, continuous measurements over at least a year's time should be made in the vicinity of fully energized lines. At the present writing, continuous measurements of this kind have not been made.

Under the existing system, the Air Quality Standards are not readily enforceable. In order to make these laws effective a review of the design and proof of intent to conform to the standards should be required before construction is started on a new installation. As we have seen, many states do not require review or licensing of electrical projects. By invoking the right of eminent domain electric companies are able to force nonconforming installations across the hills and prairie lands of America. No law requires them to demonstrate that they will not violate the new standards. No law supports the individual who raises these questions. The citizens of Ohio and many other states are not protected by the adoption of the air quality standards.

In Los Angeles, in New York, in fact in over three-fourths of the United States, Air Quality Standards are exceeded on many days during the year. When that happens a warning is issued by the local Air Pollution Board. The individual citizen must take action to protect himself. But unfortunately, there is no place to hide. Like the other watchdogs for the people, these air pollution laws are without teeth. They can contribute nothing more to the protection of our environment than furious barking while rape and robbery proceed uninterrupted.

5

A Bend in the Line

Kiziah Hostler Hough drove her rather antiquated green Chevrolet slowly down the narrow farm lane. She rattled over the cattleguard and started to turn the last sharp corner. In another minute the house would come into view, its soft pink brick framed with green foliage.

Aunt Kiz had spent a long afternoon in town. She had gone to the bank and visited the butcher shop. She had left her gardening shoes to be half-soled and had gone to three stores before finding a good serviceable cotton dress for summer. Finally, on her way out of town, she had stopped at the Cryder farm where she had arranged to buy a pair of Muscovy ducks. They were ensconced now on the back seat of her car, flopping wildly in a perforated cardboard box.

It was late and Aunt Kiz was weary. As she negotiated the last turn and saw the house set in its little grove of trees she was immediately aware that something was wrong. There seemed to be a strange empty place between the fir tree and the springhouse. I'm tired, she thought, and passed her hand over her eyes, but when she looked again the empty place was still there. It grew more obvious as she drove nearer. She stopped the car by the orchard fence and got out. There in the deep grass lay the gnarled old pear tree, the oldest fruit tree on the farm. Its trunk was sawn off six inches above the ground. A little farther on branches from the sweet cherry tree were strewn right and left; and farther still, at the back fence row, the entire clump of raspberry bushes had been chopped down, the brush tossed in a heap on either side. In the center of the newly bared space stood a stake with a red band at the top. Another stake was just visible across the pasture, almost hidden by the crest of the hill.

Aunt Kiz grasped the fence post and steadied herself. Part of her mind rejected what she saw—it couldn't be true—it must be a bad dream. But the other part knew only too well what had happened. The surveyors from the electric company had come there while she was away.

Aunt Kiz had sent the right-of-way agent packing when he had called on her eight months earlier. She had not agreed to the right-of-way nor given permission to survey. As it happened, Aunt Kiz's property would be more severely damaged by the construction of the line than any of the other local properties. The right-of-way would pass just 100 feet from the corner of her house; it would bisect the orchard, causing the destruction of at least a dozen fine fruit trees. A tower would occupy the site of the old vine-covered springhouse and rise like the skeleton of a skyscraper beside the small screened verandah that looked out across the fields.

"We'll pay you a good price," the agent had assured her. "Why don't you sell out, take your money and buy a little home in Florida? The weather down there is real good for older folks, you know, much better than here."

Aunt Kiz gave the agent a severe look. "Young man," she said, "when I am ready to be turned out to pasture, I'll pick my own time and place. I don't need the electric company to do that for me!"

The agent persisted, trying to persuade Aunt Kiz to sign a paper granting permission for the surveyors to come onto her property. "It's just a formality," he said. "It doesn't commit you on the right-of-way."

"What if I don't sign?" she wanted to know.

"Then we'll just go into the county courthouse and get a court order. When we come back here with the sheriff you'll have to let the surveyor in. It's the law."

"I don't know much about the law," said Aunt Kiz after a little thought. "But it does seem to me that signing would commit me. The Ohio Power Company is planning to rob me of my view of the fields, my springhouse, and my orchard. Signing this permit would be like giving the robbers the key to the house."

This house in which Aunt Kiz had been born and had spent almost all of her life was an unusually handsome one. It had been built early in the nineteenth century, before the first settlers realized that the native Ohio clay made very fine bricks. The bricks for this house had been brought on muleback over the mountains from Pennsylvania. Aunt Kiz's great-grandfather, who had built the house, had brought some sophisticated architectural ideas with him from the East, too. The house had a spiral staircase, a fan-lighted front door, a graceful carved mantlepiece in the living room, and wood paneling made from black walnut boards harvested on the farm. Aunt Kiz had heard her grandfather tell many times how as a small boy he had ridden the old bay horse that ground the mortar used in building the house.

Aunt Kiz had lived at the farm all her life except for a few brief years of marriage that ended when her husband was killed in the First World War. Then, after her parents had died, she continued to stay on in the old house, although she did not own it entirely. Two nephews who lived in Chicago had each inherited a one-third interest in the place. At first they had been anxious to sell and had questioned whether it was safe for her to live there alone.

"Nonsense," said Aunt Kiz, "I feel as safe here as in God's pocket."

Finally the nephews agreed to hold the house and sell off most of the farmland. Aunt Kiz lived there by herself but she filled her time with many interests. She had an extensive flower garden. She had her Sunday School class and often invited the children back to her house for a dish of homemade ice cream or a picnic lunch in her shady yard. As she grew older Aunt Kiz began hiring a village girl or boy to come in and "do" for her—to help with gardening and window-washing and the heavier housework. These young people were not paid a great deal but they were the recipients of many other benefits. She sent two of the boys through the evening course at the university extension in town. One of the girls was married in Aunt Kiz's house in front of the lovely carved mantlepiece.

Aunt Kiz's most faithful helper was "Taterbug" Brown. Taterbug was not his real name, of course, but a nickname applied first in derision by his contemporaries and later in affection by everyone who knew him. Taterbug was extraordinarily undersized. Standing less than five feet tall, he did not have an extra ounce of fat on his wiry body. His small size made it difficult for him to find work as a regular farm hand, so ever since he had left school he had worked at odd jobs. Three days a week now he helped Aunt Kiz.

Earlier this spring Taterbug had made a little duck pond in the meadow. A stream of fresh water ran down from the old springhouse past the calf-lot and disappeared into the brook that bordered Aunt Kiz's property. She had thought many times how pleasant it would be if this stream could be dammed up to make a little pond where she could keep ducks. One day she explained the idea to Taterbug.

"Why sure, that would be real nice," said Taterbug. A few hours later he had collected a wheelbarrow full of stones and had started to build a dam. Over the next two weeks the dam was completed and the spring water slowly filled the little hollow at the bottom of the meadow, making a cool sliver of blue edged with deep grass and clover.

The morning after the surveyors had visited Aunt Kiz's property Taterbug came to work, riding his bicycle. At the top of the first hill he stopped to look for the pond. There it was just beyond the calf-lot, its shiny surface reflecting the pale morning sky. As he watched a pair of big black and white Muscovy ducks climbed out on the bank and shook their feathers.

So, she done went to town yesterday, and got them ducks, he thought, and began to whistle as he rode on toward the house. He passed the orchard. Then suddenly he stopped and stood staring incredulously at the felled trees and bushes. He couldn't imagine what had happened.

After a few minutes Aunt Kiz saw him and walked out. Together they started picking up the branches, collecting them in a pile. Aunt Kiz explained the meaning of the stakes.

"Please pull the brush away and burn it, Taterbug," she said. "And saw up the pear tree for firewood. It's bad enough to lose old friends without having to look at their corpses."

Taterbug worked at this job all day with a heavy heart. He remembered the sweet raspberries that those bushes had borne, and how Aunt Kiz loved a dish of ripe berries for supper.

When he had finished clearing away all the branches and brush he was suddenly struck by the emptiness of the space that was left. He looked at the surveyors' stake and anger welled up in him. He wrenched the stake from the ground and started to hurl it into the wheat field. Then another idea struck him. He replaced the stake carefully in its original hole and ran down to the other stake in the meadow. Turning back, he lined up the two stakes, then saw on the neighboring farm a third stake, standing on a little rise of ground. Sighting along these three, Taterbug could see exactly where the transmission line would go, how it would pass straight through the springhouse and cut a swath through the center of the orchard and pass close to the side of Aunt Kiz's house. He went back to the stake he had pulled out and, muttering angrily, he moved it about fifty feet farther away from the house. There he drove it carefully into the ground again just as deep as it had been before. Once more he went back and sighted along the three stakes. Now, he noted with satisfaction, the line missed the springhouse. It even missed most of the orchard. From the side porch the tower would be partly screened by trees.

That's much nicer for Aunt Kiz, Taterbug decided. Of course, it would mean a bend in the line—but not a very big bend. He imagined just how pleased Aunt Kiz would be. Well, maybe not pleased but less downright miserable, he thought as he got on his bicycle and started pedaling slowly down the lane.

6

Trees or Towers

The representative of the Ohio Conservation Society who had attended the protest meeting made an official inquiry into the routing of the line. Would it pass through the state forest lying southeast of Laurel?

"Yes," the spokesman for the power company admitted. "But just a little of it—and, after all, what are a few trees compared to progress?"

By "progress," of course, he meant cheaper and more abundant electric power. How many trees would be sacrificed to attain that goal? Each mile of right-of-way passing through a heavily wooded area would require the destruction of about ten thousand trees.

The state forest known as Tar Hollow covers 16,000 acres. This land was purchased by the state of Ohio around 1930 and developed during the depression years, largely by WPA labor. Many miles of bridle trails and footpaths were opened up. Log shelters were built with stone fireplaces for cookouts. Several small streams were dammed to make little lakes for boating and fishing. These facilities are scattered thinly and very unobtrusively over the many acres of forest, and it is possible to spend hours in the park enjoying its beauty without encountering other people. There are protected sunny banks where trillium makes a dense carpet in the springtime and paw-paw trees drop their soft, redolent fruit in the fall. There are deep woodsy glens where Jack-in-the-pulpit and rare pink lady slippers push their

Cedric Wright

A few trees compared with progress.

Yoichi R. Okamoto

tightly curled petals through the thick mat of oak leaves and pine needles on the forest floor. In deep cool grottos streams drip over moss-covered stones and the straight trunks of beech and hemlock grow sometimes fifty feet tall before they spread their branches to catch the sunlight. From the high open ridges it is possible to look out over undulating miles of similar forest extending south and east toward the Ohio River.

Hardwood and pine forests such as these came into being over many centuries, as the result of a slow evolution from the original grasslands through a succession of different plant communities. The intense sunlight and dry soil of grasslands are poor seedbeds for most trees. Cedar and poplar seedlings are among the few that can withstand these severe conditions and even these take hold very slowly in prairie land. As these trees grow, they provide some shade so that the ground beneath them is cooler and moister, producing a favorable bed for the seeds of hardwood trees such as the oak and red maple and tulip tree. The foliage of these trees eventually shades the cedar and poplar that nursed them, robbing them of the sunlight for which they are best adapted. Now the forest floor is a better seedbed for hardwood trees than for the cedar and poplar, so the earlier species are crowded out. Eventually the towering crowns of the newcomers produce such dense shade that their own seedlings will not grow. Only certain trees—such as the hemlock, beech, and sugar maple—can sprout and mature in dense shade. Temperature, rainfall, and soil conditions create further limiting factors that determine which of these most favored species will win the struggle for survival in this particular forest. The species that thrive best in this environment finally take over to form the climax forest. No other trees can compete with them under the conditions that become established by this evolutionary process.

The successional pattern exemplified here is typical of evolutionary changes in a biosphere. The pioneer organisms exert a very strong influence on their environment, usually changing it so drastically that it becomes unsuitable for their own progeny. Transitional organisms also alter the environment, but more gradually. They eventually give way to climax organisms that are perfectly adapted to the environment that they themselves create. By maintaining a state of equilibrium with their environment, the climax organisms are able to maintain the position of dominant species with relative permanence. But if suddenly destroyed, the climax organisms cannot be immediately restored; the whole slow process must begin again.

The mature forests of Appalachia are unique in the variety of their vegetation. These forests, known to the botanist as *mixed mesophytic forests,* are characterized by a climax growth in which dominance is shared by many species: beech, tulip trees, basswood, sugar maple, sweet buckeye, chestnut, red oak, white oak, hemlock. This unusual association develops only in moist but well-drained locations in temperate climates, and on slopes protected from excessive exposure to the sun or drying winds. No other land in the Americas or even in the world (with the possible exception of certain areas of China) provides the combination of ideal conditions that supports such a luxuriant variety of climax growth.

The soil characteristic of these areas is the most productive forest land on earth. The top layer of humus, consisting of dark, almost black earth, is crumbly and friable, often extending to a depth of sixteen inches. Leaching of essential minerals from this humus layer takes place more slowly than in most other types of soil. The water and essential nutrients are retained, contributing to the rich undergrowth of ferns, flowering shrubs, and many wild flowers. The leaf litter from the undergrowth contributes to the formation of the deep humus layer.

In this ecologically balanced relationship, the soil and moisture conditions contribute to the forest and the forest cover protects and builds the humus layer and maintains the ideal moisture conditions. If any part of this association is altered, the balance may be upset too far to be corrected by natural processes. Small alterations can be corrected by nature without permanent destruction of the creative relationship, so if forests such as these are cut over once and not otherwise disturbed they often return in time to their original luxuriance. But repeated cutting over or site changes causing erosion and altering drainage patterns may completely change the character of the soil and the forest cover.

The hill region of southeastern Ohio lies in the western portion of the mixed mesophytic forest region. Many of its slopes still contain fine examples of this unusually variegated and colorful forest. Some of the area, however, has already been cut over for timber and the rationalization is often used that since it is no longer virgin forest, no great harm will be done in cutting over again. This treatment throws the ecological system so far out of balance that the mixed mesophytic forest association cannot be restored. Not only does it totally destroy a uniquely favorable forest environment, but it causes destruction that affects areas hundreds of miles away.

Protecting the Watershed

In the rugged hills of this region many streams rise and flow southward to join major tributaries, such as the Muskingum and the Scioto, which eventually empty into the Ohio River. The flow of the Ohio and its tributaries is the lifeblood of this whole part of the country. The industries, the cities, and the transportation are all vitally dependent on the river and are seriously disrupted by periods of drought or flood. A well-managed watershed helps to prevent these extremes.

The thick green mantle of forest acts as a regulating factor, damping out the rise and ebb of water flow. In winter the branches of the trees catch the snowflakes, holding them for days until they finally melt and drip onto the snow banks below. Here, partially shaded from the sunlight, the snow builds up, and melting is delayed over many weeks or months. The ground below is protected from freezing by the bank of snow and by the layers of dead leaves and other vegetation, so that as the snow melts it is immediately absorbed into the open pores of the earth. There the soft humus of the forest floor holds it like a sponge. Slowly, as the snow melts, this great underground reservoir moves down into the valleys and prairie lands where it replenishes the water table that nourishes the farm crops, the corn and wheat characteristic of this part of the country.

In the spring and sometimes in the fall, when the heavy rains occur, the many layers of vegetation in the forest absorb the rainfall and protect the soil. Branches, twigs, and leaves break the impact of driving rain so it strikes the ground gently. The deep network of root fibers holds the soil in place; and the decaying organic matter soaks up the moisture, slowly returning it at a more leisurely rate to the surrounding countryside. It is generally agreed that if the slopes of the watershed remain in timber stands, its ability to function well is automatically maintained.

It is apparent that the clearing of a broad swath across a forest creates a path which is more conducive to rapid runoff of rainfall. Along this barren strip no protecting branches shade snowbanks in winter. Shallower root systems provide less resistance to erosion. When the grasses die back in winter, the ground is more exposed to freezing and thawing and does not act effectively as a giant sponge to hold the water. The hard-frozen surface sheds the snow water like a tin roof and the straight-line routing up and down steep mountain sides provides chutes for the water to pour downhill as it floods off under the warmth of the spring sun. Thus with power saws man can destroy in a few hours an integrated system that took nature thousands of years to achieve.

In many locations today, vegetation in the rights-of-way is controlled by spraying with powerful brush killers. Spraying chemicals from a helicopter is a much cheaper way of killing the vegetation than cutting it down and removing the brush. It is also more destructive. In a cutting operation low-growing trees, bushes, and ground cover can be saved, while spraying destroys vegetation indiscriminately. The residual chemical poison seeps down into the soil and is washed by rainfall into streams which spread its destructive influence to other areas.

After seeing these corridors of death cut across our forest lands, it is ironical to read the guidelines set forth in the *Environmental Criteria for Electric Transmission Systems,* published in 1970 by the U.S. Departments of Agriculture and Interior.

> Rights-of-way should avoid heavily timbered areas, steep slopes . . . and scenic areas. . . . right-of-way-strips through sensi-

tive forest and timber areas should be cleared with curved, undulating boundaries. The notched effect of a right-of-way cross section should be avoided. . . . Careful topping and pruning of trees can contribute to this. Also small trees and plants should be used to feather back the rights-of-way from grass and shrubbery to larger trees. Rights-of-way should not cross hills and other high points at the crests. . . . The profile of the facilities should not be silhouetted against the sky. . . . Clearing shall be performed in a manner which will maximize preservation of natural beauty, conservation of natural resources, and minimize marring and scarring of the landscape. . . . Chemicals, when used, should be carefully selected to have a minimum effect on desirable indigenous plant life and selective application should be used wherever appropriate to preserve the natural environment.

The Balanced Ecosystem

Mankind is just beginning to understand the enormous complexity of the ecosystem of which he is an integral and powerfully dynamic part. The study of the intricate relationships of living organisms with their environment is a new science; it requires the development of methods quite different from those used in other scientific fields.

In most areas of scientific research, new knowledge is gained by isolating a single phenomenon and measuring it with increasing accuracy. But in the study of living things it is impossible to isolate a single phenomenon. The complex mechanisms taking place within a single living cell must be understood as integral parts of a dynamic process involving many interdependent activities and cyclic patterns in time.

Biologists attempting to understand these complex interrelationships are hampered by the tools of thought carried over from the older sciences. The public is also confused and frustrated in dealing with this new science. One day they read that scientists are concerned that mankind is altering the climate of the earth, the next day that chemists have shown that such changes are "negligible."

Seen in the light of the old isolated-phenomena view of science, it seems reasonable to assume that a small addition to a process that occurs naturally cannot be harmful. Nature will take care of it. It is all right to add more ionizing radiation to the environment because we are constantly exposed to background radiation. It is all right to add heat to the lakes and rivers because the amount added is only a small percentage of the natural heating that occurs every summer. It is all right to generate more ozone because we are always exposed to some, anyway. But these assumptions fail to take into account the delicate equilibrium maintained by nature.

Over the millennia, the variable factors present in each biological community have been balanced out to achieve a relatively integrated and stable system. Organisms evolved that were able to handle the usual changes in variable elements in their environment. Accelerate one variable in this dynamic relationship even a small amount and the compensating factors are not sufficient to maintain the equilibrium. Nature, given enough time, might achieve a new balanced system, but the time scale for this kind of natural evolutionary development is in hundreds of thousands of years.

To visualize the ecological relationship one might imagine a teeter-totter with many arms, all so perfectly balanced that the children riding it are suspended motionless in space. If one child moves forward just a little bit the delicate balance is disturbed. The change in weight distribution is only a very small percentage of the total forces involved but, in the absence of any compensating change, the entire system is upset and a drastic disturbance results. In order to make a firm prediction about the significance of altering one element in an existing ecosystem, all of the intricately interwoven processes must be identified, the rates must be known very accurately, and so must the manner in which they act upon each other. One of the famous unsolved problems in physics is the calculation of the forces produced by three gravitating bodies acting on each other. The mathematics of this relationship have never been precisely solved. But compared to the interrelated forces in an ecosystem, the three-body problem is simplicity itself.

Perhaps someday with the aid of computers and mathematical models, a quantitative understanding of these reactions may be achieved;

but today the best answer is just an educated guess. At the present state of our knowledge, conservation of the environment can best be achieved by a conservative attitude toward any change initiated by man. Any change that appears to be progressive should be questioned even if it is very small. And ecological damage is most apt to occur in cases where several new factors are working simultaneously in the same direction.

The Role of Forests in the Ecosystem

Forests cover about a tenth of the earth's surface and fix almost half of the biosphere's total energy. Through the process of photosynthesis they combine solar radiation with carbon dioxide and water to form the organic compounds that are the primary source of energy for all living things. Studies at Brookhaven National Laboratory have shown that one square yard of oak-pine forest produces on the average 2.1 pounds of new organic matter each year. The largest portion of this increase is, of course, in the form of new timber. The destruction of a 200-foot strip 100 miles long would mean a loss of 27,000,000 pounds of timber and other organic matter every year; and this organic matter is the ultimate source of fossil fuel and electric power.

The forests of the world are the main consumers of the carbon dioxide in the atmosphere. They and the fossil fuel deposits are the principal reservoirs of biologically fixed carbon and energy stored in organic compounds. The areas of the earth's surface that are capable of maintaining highly productive forest cover are limited; deserts and arctic tundra cannot contribute much to the total fixation of carbon by photosynthesis.

As the trees absorb carbon dioxide from the air and synthesize carbohydrates, they perform another function vital to life: they release molecular oxygen as the end product of photosynthesis. The amount of free oxygen released is directly proportional to the amount of organic matter created; when the plant dies, this organic matter is decomposed or burned by oxidation processes which use up the same amount of oxygen that the plant has released during its growth. This cycle of release, storage, and utilization of oxygen is one of several basic cyclic processes that depend upon the total amount of photosynthesis that has taken place during the earth's history. The amount of free oxygen and carbon dioxide in the atmosphere is regulated by the amount of carbon compounds stored in living organisms and in the deposits of ancient organic materials such as coal, gas, and oil. These fossil fuels are concentrated sources of energy. During the past century large quantities of them have been burned to produce heat and electric energy, and there is considerable concern among scientists that carbon dioxide is accumulating too fast in our atmosphere as a result. Since 1860 there has been a 10 per cent increase in carbon dioxide concentration.

An increase in the level of carbon dioxide is believed to cause a temperature rise due to the so-called greenhouse effect. Carbon dioxide molecules in the atmosphere (like the oxygen and nitrogen molecules) allow the energy of the incident sunlight to pass through to the earth's surface. However, when this solar radiation has been acted upon by vegetation and other matter, it is converted to the longer waves of heat energy; carbon dioxide molecules absorb and reflect these wavelengths so that the energy does not pass back out into space. If fossil-fuel consumption increases at the present rate, by the year 2020 the increase in carbon dioxide concentration would, *if it were an isolated phenomenon,* cause an increase of 8 to 10 degrees Fahrenheit above 1950 levels in the earth's mean temperature. However, there are many related processes also occurring. As fossil fuels are burned, dust and fine particles accumulate in the atmosphere. These particles have the effect of screening out some of the solar radiation *before* it reaches the earth's surface. The carbon dioxide concentration also affects the rate of photosynthesis; so does the ozone content; and again, all these factors are interdependent. The complexities are not entirely understood. In the recent technical literature scientific arguments rage back and forth concerning the ultimate effect of these factors. I think it is fair to say that no one is sure exactly what net effect these changes will have on the earth's climate and oxygen supply.

We do know, however, that changes have been occurring. Records have shown that over the past century there was a long period of rising temperatures. Then this warming effect decreased and lately there has been a small but consistent decline in the earth's mean temperature. It is generally believed that the rise was due to increasing carbon dioxide and that the reversal has been caused by larger accumulation of particulates in the air.

In the meantime, carbon dioxide levels continue to rise. If, as environmentalists hope, we succeed in reducing the amount of dust and fly-ash in the atmosphere, then the rise in temperature which has been masked by the screening action of the particulates would suddenly make itself felt. Important climatic changes could occur, causing melting of glaciers and flooding of coastal areas. Although still in the realm of guesswork, these possibilities deserve serious thought, especially since several of these changes are working in the same direction—toward increasing the carbon dioxide content of the earth's atmosphere.

The public utilities are burning increasing amounts of fossil fuel to produce more electricity; this burning creates higher levels of carbon dioxide in the air. At the same time, they are building overhead transmission lines that require the destruction of hundreds of thousands of trees. Again the carbon dioxide level will be increased. Near the rights-of-way damage to vegetation by erosion and by chemicals sprayed from helicopters and created by corona discharge will further reduce the forest growth. These effects all add up and are, therefore, particularly dangerous as a threat to the balanced economy of nature.

An increasing fraction of the total energy fixed throughout the earth's history is being diverted to the direct support of man. Simultaneously, the amount of vegetation that fixes this energy is being continuously reduced. The trend is progressive and the resources consumed are irreplaceable.

A living ecosystem, said Paul Sears, must be seen as a manifestation of the flow of energy. "The health of a landscape, that is, its capacity to sustain life, is measured by the efficiency with which use is made of solar energy between its reception and its inevitable dissipation. A truly healthy landscape . . . is the rule rather than the exception in nature. It represents what the physicist calls an open steady state, a condition of equilibrium that continues to receive energy, do work and at the same time keep itself in working condition. . . . One of the surest and readiest means to diagnose the health of a landscape is its esthetic quality. Ugliness is an almost certain symptom of inefficiency, beauty an equally valuable sign of harmonious adjustment."

The erection of high-voltage transmission lines across forested areas destroys many acres of precious woodland. It replaces hundreds of thousands of graceful green trees with stark steel towers. It blocks out the gentle night noises of the forest with the loud ominous hum of nearly a million volts zinging overhead and fills the fresh air of damp woodland places with a brew of highly reactive chemicals.

Already in many places across our land, steel forests are replacing the delicately interwoven pattern of the living forest. This many-layered gossamer web of green that clothes the mountains and hillsides of our beautiful planet is the primary energy producer of the biosphere, the purifier and regulator of the essential ingredients in our atmosphere, as well as a continuing source of recreation and inspiration. The laceration of this complex living system with broad slashes of destruction in order to achieve cheaper electric power can hardly be equated in any meaningful way with progress.

Hopefully, the time will come when progress will be equated with an improvement in the quality and diversity of life rather than with the quantity or cheapness of things. When that time comes it will seem incredible to our descendents inhabiting an impoverished earth that we could have destroyed in the name of progress thousands of acres of these forests that are such an outstanding example of the incredible diversity and richness of nature.

7

David and Goliath

In the general store in Laurel, Irma Kellenberger began to sort the day's mail. The mailboxes were little glass-fronted partitions in a high wall of golden oak. Along the top of the wall above the boxes a dozen "Wanted by the FBI" notices had been tacked and hung at crazy angles.

Irma knew the location of each box by heart. She had been sorting the mail into these same boxes for more than thirty years. The Jacobs had a postcard from their son in Germany. And the Helms finally had a letter from their boy in Vietnam. Again today, the Muellers had more mail than anyone else. There was one letter from the Federal Power Commission in Washington and one from a state senator. But Irma noticed that both the letters were one sheet thin and obviously contained only a line or two of type. Two more turn-offs, she thought. In the month that had passed since the protest meeting in the schoolhouse, letters had poured into and out of the Laurel postoffice. Many of them bore important-looking return addresses but none had produced any assistance in the protest against the transmission line.

Across the store, Donald was refilling the penny gum machine and sorting the packages of socks and underwear on the "clothes bar." Then he laid the pile of Columbus papers on the oak counter. They arrived a day late in Laurel but there were several families who ordered them regularly in spite of the delay. He glanced at the front page and then flipped over to the second section. Large headlines jumped out at him:

POWER LINE ROUTE
STIRS PUBLIC IRE

A picture showed several local residents silhouetted against a view of Tar Hollow State Forest.

"Say, look here, Irma," he called out. "We're in the *Dispatch*!"

Irma put down the pile of unsorted mail and walked over to the counter. The article, written by the "outdoor editor," pointed out that the right-of-way through public land had been granted without public hearings. However, the author explained, Ohio law does not require such hearings. He also described in some detail the protest group that had been organized around Laurel and had brought the matter to public attention.

The bell on the front door dinged as Rose Wolf and Mildred Freeman came in for their morning mail. A minute later, Glenn Pontius drove up in his farm truck, parked in front, and entered the store. He was followed by Ovid Helm and Damon Orr. Soon there was a little group around the counter, reading the article out loud and savoring every phrase: "The granting of this right-of-way is a violation of public trust. . . . It is a gross inadequacy of Ohio law that public hearings are not required in such matters."

"How do you like that—that's tellin' em."

The bell dinged again and Alice Mueller entered. It was Saturday, Alice's day off.

"Say, Alice, you seen this spread in the *Dispatch*?"

Although she had not seen the article yet, Alice knew all about it. Her friend Howard Smith of the Audubon Society had attended the protest meeting and had given the story to the outdoor editor. They had both discussed the details with Alice this past week.

"The Ohio Power Company saw the article," she said, laughing. "I got a real interesting call from them this morning. They think it may be possible, after all, to avoid running their line through Tar Hollow!"

"Well, ain't that great!"

"I guess it takes a birdwatcher to flush those buzzards out of their cover!" exclaimed Ovid Helm. "Sure do hate publicity, don't they?"

Like David of the Biblical story, the birdwatchers of Whistler Road had found a small chink in the armor of the Philistines. As long as freedom of the press is preserved, publicity offers one possible way of fighting the giants of industrial power. They know that public opinion, if mobilized in time, may make it difficult for them to impose their plans upon people.

The day after the newspaper article appeared the power line was rerouted around Tar Hollow. The altered route, however, still passes through miles of heavily timbered hill country in Ross and Vinton counties, where, in direct defiance of the environmental criteria set forth by the Departments of Agriculture and the Interior, the power company plans to carve broad straight swaths through scenic forests, crossing hills at their crests and displaying the ugly profiles of metal scaffolding against the sky.

8
Little Tranquilizing Pills

In response to requests for information concerning the design of high-voltage equipment and its environmental impact, power company officials offer opinions and what Rachel Carson called "little tranquilizing pills of half-truth." "We urgently need an end to these assurances," she said, "to the sugar coating of unpalatable facts. It is the public that is being asked to assume the risks. . . . The public must decide whether it wishes to continue on its present road, and it can do so only when in full possession of the facts."

Today it is just as hard to get at the truth as it was in 1962 when she wrote these words, and the risks to which the public is exposed have increased enormously. Rachel Carson put her finger on the principal reason why facts are withheld from the public. When people are in full possession of the facts, they can make their own decisions on the proper course of action. But the industries concerned do not want the people to make such decisions. Instead they seek general acceptance of their own decisions. This can best be achieved by statements that contain value judgments rather than information.

In answer to questions concerning the amount of electrochemical oxidants produced by corona discharge, power company executives answer: "The amounts are minute and rapidly dissipated." This statement does not convey any information on which to base an appraisal of the safety of the line.

Very small amounts of these chemicals—less than five parts per hundred million—can be damaging to living things. Is five parts per hundred million "minute"? Some people would consider one part per million minute; and that concentration is extremely dangerous.

Damage to the Whole Organism

Each individual organism is a beautifully integrated system consisting of hundreds of millions of finely adjusted components. These components are related in a dynamic manner involving the flow of tiny energy pulses from one part of the system to another, causing the growth, the responses, the activity that we associate with life. It is easy to see how the exposure of such delicately balanced systems to the little bullets of extra energy carried by ozone and the other related oxidants causes destruction to molecules and interference with many biological processes.

Although various types of oxidant injury have been observed for the past thirty years, the cause was not immediately identified. As early as the 1940's tobacco growers in the vicinity of Washington, D.C., began complaining of damage to their crops. The leaves of the tobacco plants were becoming mottled with brown spots. First slightly water-soaked or bruised-looking areas appeared on the undersurface of the leaves. Then these areas dried out, leaving bronze-colored spots of dead organic material. Soon similar reports came in from other parts of the country. In California this "stippling" was occurring on grape leaves, and the yield from the vineyards was reduced. Cucumber plants in New Jersey were flecked with brown, and onion plants in Wisconsin were dying back at the tips. In the San Bernadino and San Jacinto mountains near Los Angeles, hundred-foot ponderosa pines began turning yellow, and all through the eastern states reports came in of white pine showing the same symptom.

The incidence of these attacks was erratic. At first they were diagnosed as fungus diseases or insect invasions. It was only after the recognition that ozone and other oxidants were important factors in air pollution that the true nature of these attacks was discovered.

It is now believed that ozone is the primary cause of injury. PAN also contributes to the damage and is especially effective in attacking young leaves. Extremely low concentrations of PAN, as little as ten parts per billion, can cause injury to sensitive plants. Ozone usually attacks leaves in the more mature stages of growth. Many studies now show that visible leaf injury to sensitive plants is caused by exposures to concentrations of ozone as small as .05 ppm (parts per million) for an eight-hour period. This is only twice the normal unpolluted background concentration.

Plants vary considerably in their sensitivity to ozone. Tobacco, petunias, and bluegrass are especially sensitive. Ponderosa pine and citrus fruit trees, on the other hand, are relatively resistant. These trees show visible damage when exposed to concentrations of about .2 ppm for eight hours.

Even before visible injury appears on the leaves, however, damage has occurred which affects the growth, yield, and life expectancy of the plant. During the last ten years, orange and lemon trees near Los Angeles have been losing their leaves early, producing smaller fruit, and achieving poorer growth. Consequently the raising of citrus fruit, which was a major resource for this area, has experienced a serious decline. Growers of carnations and petunias in many parts of the country have reported significant reduction in the number of flowers as well as a reduction in size. Radish crops have declined as much as 50 per cent in yield. Such gross effects must reflect a serious disturbance in the physiological processes of the plants.

Under controlled greenhouse conditions, greater understanding has been gained of the nature of the oxidant's interference with vital processes. Plants exposed to low levels of oxidant for long periods of time show pale-green or yellow leaves—color changes associated with a reduction in chlorophyll. This remarkable green chemical, which enables the plant to utilize the energy of sunlight in photosynthesis, is contained in a layer of the leaf called the *palisade* cells. And it is in this layer that the damage from oxidant occurs.

Several scientific studies have shown that the rate of photosynthesis is very markedly reduced following exposure to ozone. At the School of Forestry at Yale University, Dr. William Smith and Dr. Daniel Botkin have been measuring the photosynthetic rates of white pine saplings. Exposure of branches to consecutive doses of .5 to .8 ppm for 3 hours was sufficient to reduce photosynthetic rates by approximately 80 per cent. They conclude that damage to non-woody vegetation is likely to occur whenever the ozone reaches or exceeds .05 ppm for one hour or more. They estimate that during the growing season this dosage occurs, on the average, one out of every four days throughout the entire state of Connecticut.

Recent measurements of oxidant levels in rural areas of West Virginia and Maryland, far from any urban centers, showed *average* hourly oxidant concentrations of .05 ppm over a four-month period. Throughout this whole region pine trees are showing chlorotic damage and commercial tree growers have suffered disastrous losses.

It is apparent that the oxidant level which can be tolerated by vegetation has already been exceeded in many areas of the United States, in rural districts as well as urban. Any factor that increases this level will add to the present amount of destruction of crops, forests, gardens, and lawns all over America. Further damage cannot help but reduce the quantity of food raised in this country, at a time when increasing world population is making excessive demands upon the food supply. It will represent financial losses to the farmer and truck gardener. Perhaps most important, it will decrease the total amount of photosynthesis over the earth. All animal life is completely dependent upon this remarkable process, which converts the energy of sunshine into organic material, storing the energy in a form that can be used as food by animals, and, as a by-product, releases oxygen into the air. Our supplies of food, our stores of oil, coal, and gas, all derive from this process. Scientists believe that the green cover of vegetation is the primary source of the oxygen in the earth's atmosphere. There is no atmospheric oxygen on the moon or on Venus; there appears to be no life on these planets either. The thin film of living matter on earth really relies entirely on the magic of chlorophyll reacting with sunlight. To poison the operation of this delicate biochemical process, which is still only dimly understood, is truly to tamper with one of the mainsprings of life.

Laboratory research aimed at discovering the degree of toxicity of ozone to human beings is hampered by the fact that few people are willing to volunteer for such experiments. The small number of tests that have been conducted on volunteer subjects have used, of course, quite low levels of ozone. Involuntary exposure, as to Los Angeles smog, has occurred for too short a time and is not sufficiently documented to provide the necessary data. In addition, the fact that the human lifetime is relatively long makes experimentation with human beings very unsuitable when quick answers are needed.

These difficulties can be circumvented by using small laboratory animals. Experiments with mice and guinea pigs provide clues that suggest the implications to human health. Of course, the results of experiments with animals must be applied with caution to human beings. The exact rates of certain reactions, levels of sensitivity, and so on cannot be directly transferred to people. However, a great deal of valid information has been learned concerning the general nature of the toxic effects and the type of reactions that occur. The few experiments done on human volunteers seem to produce the same pattern of responses as the animal studies.

When sensitive laboratory animals, like rats, are exposed to large doses of ozone (about 6 ppm for four hours), 50 per cent die from shock, with massive swelling of the lung tissue and hemorrhage. Smaller doses (.1 ppm), repeated daily for a year, cause lung tissues to undergo permanent changes similar to those found in emphysema and fibrosis. Even shorter exposures—one hour once a week for 52 weeks—result in signs of accelerated aging. Concentrations of ozone as low as .15 ppm for 3 hours have been reported to reduce significantly the animal's resistance to bacteria, particularly those causing respiratory infections. As the scientists tersely phrase it, "Mortality is enhanced . . . by ozone exposure."

The animal experiments to date suggest that ozone exerts its toxic effect primarily on the respiratory system; and the few experiments on humans confirm this opinion. Subjects showed decreased vital capacity in the lungs, accompanied by chest pain and cough. Drowsiness, headache, and inability to concentrate were also reported after only brief exposures. High air temperatures and high humidity increase these symptoms. Exercise or physical labor also greatly enhance the effects.

In September 1971 an incident occurred which brought the dangers of oxidant pollution to national attention. The players on the high school football team in Quibbletown, New Jersey, were stricken with a strange malady. During routine afternoon practice one day they complained of sharp pains when they inhaled, a rawness in their throats, and an inability to get their breath (as though they had been punched in the chest). As the condition progressed it included vomiting and weakness, and in some cases a tingling sensation in the arms and legs. These boys were aged 12 to 14, athletic, and in prime physical condition. The air that day in the New Jersey area had a slightly grayish-yellow cast, but it did not look heavily polluted. The air pollutant factors routinely reported were at normal levels. Health authorities finally diagnosed their ailment as respiratory damage caused by air containing elevated oxidant pollution, about .08 to .09 parts per million. In spite of incidents of this kind, demonstrating the health significance of oxidants in the atmosphere, measurements for total oxidant levels are made in only a few locations and are not usually reported in the air pollution indices.

A number of commercial uses of ozone were instituted—in fact, inflicted on the public—before adequate tests had been made of their effectiveness and safety. It has been known since early in this century that ozone suppresses the growth of microorganisms. This germicidal property was employed to suppress fungi and bacteria growth associated with food spoilage, to purify drinking water, and to treat sewage. For a number of years ozone was deliberately added to air-conditioning and ventilating systems in public buildings. Little ultraviolet bulbs that generated ozone were installed in clothes driers to impart a fresh sterile smell to laundry. Similar bulbs were used in public washrooms to sterilize toilets. Frozen food lockers add ozone to the air to suppress spoilage. And in hospitals ozonizers are often placed beside the beds of terminal cancer patients to destroy the offensive odor caused by that disease.

Many of these uses are now known to be health hazards. Research results coming in slowly over the years and reported by the U.S. Department of Health, Education, and Welfare have shown that *ozone is not effective as a bactericide unless it is used in concentrations which are toxic to human beings* (about .04 ppm).

The germicidal effectiveness of ozone varies with its concentration, the relative humidity, and the species of bacteria. High relative humidity makes ozone much more lethal as a germicide; under dry conditions, bacteria are quite resistant to it. One researcher found that concentrations which killed dry typhus bacilli, staphylococci, or streptococci in the course of several hours, killed guinea pigs first.

In the light of these findings, a few of the commercial applications mentioned above have recently been curtailed. The use of ultraviolet bulbs to sterilize public toilets has been discontinued. Ozone is no longer deliberately added to air-conditioning and ventilating systems in public buildings. However, it is generated as a by-product in the electrostatic precipitators used in many installations to remove dust and other particulates. These precipitators operate by ionizing the air and the suspended pollutant particles by means of high-voltage electric discharge. This is essentially the same process that is responsible for ozone formation by corona on high-tension lines.

Efforts have been made to suppress ozone formation in precipitators designed for systems recirculating air inside buildings, but as recently as 1969 acceptable levels for industrial installations were set at .1 ppm. Exposure to this concentration for eight hours a day is now known to be biologically damaging; and many installations made before 1969 are in operation today.

Thus commercial applications resulting in high levels of ozone still persist in spite of unequivocal knowledge of the danger of such exposure, showing again how difficult it is to change a technology after it has become entrenched. These cases bring home to us once more the need for adequate testing of every technological innovation *before* it is inflicted on the public.

It is possible that when all the results are in, ozone will prove to be a more efficient sterilizer of plants, animals, and men than it is of bacteria. Studies with tobacco plants show a 40 per cent reduction in pollen germination following a 5- to 24-hour exposure to .1 ppm of ozone, representing a major loss in reproductive capacity. In several experiments the occurrence of sterility in mice was doubled by chronic exposure to oxidant concentrations of from .1 to .5 ppm. Those animals that did produce young had smaller litters, with a higher incidence of neonatal death and defective offspring. The scientists reporting these experiments suggest that these effects may be caused by the oxidant altering the genetic composition of the sperm.

Misprints in the Genetic Code

Every organism, from a bacterium to an elephant, starts life as a single cell containing a complex organic molecule, DNA (short for deoxyribonucleic acid), which carries the complete instructions for the growth and form of the organism. Although the DNA molecule is so small that you could put one from every human being on earth into a thimble, it is so fantastically complicated that computers are needed to help biochemists work out the interrelationships of its various parts. These parts are joined together by bonds that can be broken by the sudden application of energy. X-rays have been found to destroy these bonds; so does ultraviolet light; and so do certain chemicals which carry extra energy. The dislodged part may subsequently recombine with the parent molecule, but in so doing its relative position in the structure may be changed.

In the DNA molecule it is the arrangement of parts that determines the genetic code, just as a printer setting type for a book uses a small number of symbols in a vast variety of combinations and can create an infinite number of texts. If one letter is knocked off and put back in the wrong place—for instance if *art* were changed to *tar*—the whole meaning of the sentence would be changed. In a DNA molecule a misprint of this kind results in a change in the genetic instructions and is known as a mutation.

Now remember that ozone and the related oxidants are characterized by the possession of extra energy. The molecules or molecular fragments enter the body through the lungs or digestive system carrying their little packets of energy. Once inside they enter into chemical reactions with organic molecules in a number of different ways. If they encounter a DNA molecule, the extra energy they release can break one of the delicate bonds and a mutation may result. Since these mutations are random and violent events at the molecular level, the results are almost always detrimental. (You could hardly expect to do much good by starting a shooting match in a china shop). If the altered code occurs in a reproductive cell it will be replicated each time the DNA molecule reproduces itself, and the new organism will turn out a little different from its parent. A rough rule of thumb is that ninety-nine out of a hundred mutations produce less "fit" individuals—the albino, the hemophiliac, the mongoloid, the dwarf are all examples of fairly common mutations in human beings. But there are a vast number of lesser-known abnormalities caused by mutation and many do not show up until the reproductive cells (which carry the DNA molecules) have combined in the second or third generations.

Each photochemical oxidant molecule has a certain chance of producing a mutation when it enters into reaction with a DNA molecule, so if *any* of these oxidant molecules are present, there is some chance of mutation. This is what is meant by saying that there is no "threshold" for the effect. Biologists and chemists who have experimented for years with these phenomena are quite well agreed that there appears to be no safe dosage below which mutation will never occur. It may result from a single bullet of energy striking the DNA molecule.

The reader may have recognized by now a similarity between the biological effects of these oxidants and the effects of radioactivity. The oxidants we have been considering are described as *radiomimetic* because they imitate radioactivity. The mechanism in the two cases is very similar. Radioactive substances emit high-speed electrons, pulses of energy, and molecular fragments, which interact with molecules to produce ions. The ions, free electrons, and free radicals carry their disrupting influence into the surrounding medium. When they encounter living matter the packets of energy that they carry provoke violent reactions which disrupt the delicate harmony of the living cell. Here again, biochemists have found no evidence that there is a true threshold of dosage below which atomic radiation produces no harmful effects. Even at low dosages the effects are appreciable. Ozone and the closely related oxidants are radiomimetic and have no threshold for biological effect. "Theoretically," says the Committee for Community Air Quality, "the recommended air limit for O_3 and related oxidants should be zero, or as close to zero as possible."

Carcinogenic Properties

There has been a marked increase in malignant disease in this country over the past few decades. In 1960 there were 279 new cancer cases reported for every 100,000 population. The statistics compiled by the American Cancer Society show that, ten years later, the incidence of the disease had increased much more rapidly than the population. In 1970, 314 new cases were reported for every 100,000 population.

Even though the mechanisms that trigger cancer are still not entirely understood, some facts are known and should be kept clearly in mind:

1. Exposure to certain chemicals, X-rays, and radioactivity increases the likelihood of an individual's developing cancer. Many tests and statistical studies have demonstrated that this is true beyond any reasonable doubt.

2. The prevalence of these cancer-causing agents is steadily increasing in our environment. Insecticides, food additives, drugs, air pollution, X-rays, fallout from nuclear testing, and radioactive waste from power plants—each of these factors may contribute only a small amount in itself, but added together they represent a very significant additional exposure of the whole population.

Several members of the oxidant family, as we have already noted, are believed to be cancer-causing agents. Ozone entering the organism through the lungs or digestive tract can react with organic molecules to produce singlet oxygen, which attacks DNA and enzymes in ways that may trigger malignant growth. Free radicals are also thought to be involved in some types of cancer.

In 1953, Dr. Stephen Zamenhof, a biochemist at Columbia University, demonstrated that nitrous acid attacks DNA, garbling a part of the genetic code. By causing mutations in the digestive tract, nitrous acid may be one of the agents that cause cancer of the stomach. Studies have shown a correlation between high levels of air pollution and mortality rates due to cancer. In Buffalo, researchers found that the death rate from gastric cancer was considerably higher in regions with the most polluted air. And a British study found that the death rate from lung cancer was two to three times greater in large towns than in rural areas, where the air was cleaner.

Admittedly, these relationships are subject to further verification. It may take scientists another twenty years or more to establish firm proof based on experimental tests. In the meantime, thousands of people may die from these causes. In the presence of a considerable body of evidence pointing to the carcinogenic properties of electrochemical oxidants, it seems only prudent and moral to protect the public against deliberate and unnecessary exposure. Probably most people would rather die of old age than of cancer.

Fountain of Old Age

The experiments that have been conducted on exposure of all different types of living things to photochemical oxidants have turned up one consistent result: these oxidants speed up the aging process.

Chronic exposure to low levels of photochemical oxidants causes vegetation to pass through the maturing and aging cycle more rapidly than normal. Exposed trees change to autumn colors too early and drop their leaves before their time. "The pattern," notes the report of the U.S. Department of Health, Education, and Welfare, "is usually not distinctive, appearing only as an early senescence."

Studies of the effect of inhaling ozone on the red blood cells in both animals and human volunteers "indicated an acceleration in the aging of the cells." Lung tissue from exposed animals showed fibrosis and hardening typical of changes that normally occur with old age.

Dr. H. E. Stokinger reported that, in experiments demonstrating accelerated or premature aging in rabbits after one year of weekly one-hour exposures to ozone, the animals developed premature hardening of the cartilage, coarsening and "unthrifty" appearance of the fur, severe depletion of body fat, and general signs of old age such as dull eyes, sagging eyelids, and decreased activity. This description strikes too close to home; the process of aging is one that we would prefer not to have accelerated.

Biologists are just beginning to get some clues to the causes of senescence. One of the most recent theories is that aging is brought on by random destructive reactions by free radicals in the body. Once inside an organism, radicals can initiate complex chain reactions which magnify their effect enormously. As errors gradually accumulate the body cells begin to malfunction and eventually to die. Damage occurring over the lifetime of the organism causes the slow death that we call growing old.

Another type of chemical damage, described by Dr. William A. Pryor, results in changes in connective tissue such as collagen. The biological role of collagen depends on its high plasticity and its ability to bear stress and maintain shape and form. It is present throughout the body but occurs in particularly high concentrations in flexible organs such as the lungs, blood vessels, skin, and muscles. With age collagen fibers become denser, stiffer, thicker, and less plastic. Similar changes occur as a direct consequence of exposure to ozone. It has been observed for a long time that rubber becomes brittle and cracks when exposed to atmospheres containing ozone and that this is caused by changes in the large molecules that give rubber its elasticity. When we compare the soft resilient skin and flexibility of youth with the sag and brittleness of age, it is hard to avoid the conviction that just such a process must be occurring year after year in the human body.

Perhaps some day scientists will find a way of slowing down and counteracting these chemical changes. Like Ponce de Leon, we would all like to discover a fountain of youth. In the meantime the concentrations of photochemical oxidants and free radicals are building up all the time in our atmosphere; the sources of these very reactive substances may prove to be fountains of old age.

In spite of the fact that corona discharge is known to be an efficient generator of these very chemicals that are biologically damaging in concentrations as low as ten parts per billion, and in spite of experimental evidence that any addition to the already high levels of these pollutants may cause mutations, malignant disease, and accelerated senescence, electric companies answer questions from concerned citizens with statements that the amounts produced are "minute and rapidly dissipated."

Patterns of Distribution

The term "dissipation" implies both dispersion and disappearance, but it is a well-known principle that neither matter nor energy can truly disappear. It is never destroyed but only converted to other forms. Chemicals broadcast on the winds and waters of the earth are redistributed but not dissipated. The patterns of redistribution are complex and difficult to predict. It is a rare situation indeed when redistribution results in uniform dilution.

For example, a surprising distribution pattern of total oxidant concentration has been discovered in the Los Angeles area, where many readings have been collected over a number of years. The surface winds in the Los Angeles basin blow predominantly east off the ocean in the spring, summer, and fall months. In October 1965, concentrations of oxidant were recorded at four different sites: West Los Angeles, Los Angeles itself, Azusa, about 20 miles east of the city, and Riverside, about 30 miles east of Azusa. Graphs of these measurements show that peak concentrations occurred at about 11 A.M. in West Los Angeles, noon in Los Angeles, 2 P.M. in Azusa, and 4 P.M. in Riverside. The time difference was about what would be expected from the direction and speed of the prevailing winds. However, the peak concentrations recorded in Azusa and Riverside were approxi-

Cedric Wright

The winds and waters of the earth move in complex patterns. Chemicals broadcast upon them are redistributed but not dissipated.

Environmental Protection Agency

mately as high as those recorded in Los Angeles. The reasons for this lack of dilution as the pollution moved eastward are not clear. Some factor is counteracting the dispersal that one would normally expect from redistribution by the wind. As the oxidants move eastward, synergistic chemical reactions may cause oxidants to be formed at a rate equal to the rate of dilution. Local pollution conditions may add to the oxidant levels or wind patterns may tend to funnel the pollutant.

Temperature Inversion

Dispersal of pollutants into the atmosphere is strongly influenced by local weather and the way temperature varies with height above the earth's surface. Normally, the temperature of the atmosphere decreases with height for several miles above the earth. In the daytime the ground is warmed by the sun. Heat is transferred to the air near the ground and this warm air, being lighter than cold air, rises, diffusing into the atmosphere at higher altitudes. However, at night the ground cools more rapidly than the atmosphere so that the air at the earth's surface cools off first; this heavier layer remains near the earth's surface, creating a blanket of stagnant surface air that is often laden with pollution. This typical nighttime temperature inversion occurs on more than half of the nights throughout most of the United States.

When the sun rises and warms the earth the inversion patterns usually begin to dissipate. Surface winds also help to break up temperature inversion conditions. However, there are regions where surface winds are blocked by mountain ranges or where winds are normally light because of the typical circulation patterns of jet streams and trade winds around the globe. In these localities temperature inversions may persist and build up for days. The Los Angeles basin is one of these regions; this contributes to its severe air pollution problems.

Occasionally barometric pressure systems cooperate with the other factors to cause a very widespread and long-lasting temperature inversion pattern. Episodes of this kind can constitute serious health hazards. There have been a number of famous air pollution disasters in which many people died, such as the episode in the Meuse Valley in Belgium in 1930 and the one in Donora, Pennsylvania in 1948.

Since 1960 the Environmental Sciences Administration has been issuing warnings when widespread high air pollution was expected. Records have been kept and a number assigned to each episode of High Air Pollution Potential (HAPP). Many of these episodes have lasted for several days and have affected a large part of the country.

In late August 1969, Episode 104 affected twenty-two states east of the Mississippi. For ten days High Air Pollution Potential warnings remained in effect for a large area from Indianapolis southeast across Ohio and into West Virginia. Most of Kentucky and part of Tennessee were under the alert for seven to eight days. Episode 104 affected more than twenty million people.

Measurements in many localities and under varying wind conditions are needed in order to understand the distribution patterns of airborne pollution. One fact, however, appears clear. It is not safe to assume that oxidants or other pollutants will be dissipated by the wind. This planet of ours is a small and tightly bound ecological system. Lead expelled in automobile exhaust in Chicago or Los Angeles turns up in glaciers in Greenland. DDT sprayed over farmlands in Illinois and California reappears in fish in the Arctic circle. Far from being dissipated, these elements are present in alarming concentrations thousands of miles from the original source.

Selectivity and Food Chains

One of the reasons for uneven distribution patterns lies in the nature of chemical and biological processes. Even simple inorganic chemical reactions are selective. Stir a spoonful of vinegar into a glass of water and the vinegar will become evenly distributed throughout the water. But add a spoonful of olive oil to a glass of water and the oil will remain in a concentrated layer on top of the water. Thus the degree of dispersion depends upon the other elements present in that portion of the environment. In some situations a chemical may be concentrated rather than dispersed.

Living organisms have a remarkable ability to select and concentrate chemical substances. Each living system absorbs some elements from its environment and rejects others. As the organism grows it accumulates a high proportion of the selected elements. This organism and others like it then serve as a food source for larger animals which absorb and concentrate the element even further. When these animals serve in their turn as food for man, the element may be stored in the human system in volume densities several thousand times greater than its original density in the environment. Striking examples of this phenomenon of biological concentration have come to light in recent years.

When atomic weapons were tested in Nevada and the islands of the South Pacific in the 1950's, a number of radioactive elements were ejected into the atmosphere, among them two long-lived products, cesium 137 and strontium 90. These elements are dispersed by the winds and—one might suppose—should have been quickly distributed throughout the earth's atmosphere. However, it turned out that prevailing wind patterns tended to create layers of higher density of these elements at certain latitudes. Eventually, the radioactive molecules were washed down to the earth's surface by rain and snow. Thus patterns of precipitation also played a part in distributing a higher proportion of cesium 137 and strontium 90 onto certain geographical regions.

Where these elements collected on soil and vegetation, they were absorbed in varying degrees depending upon the chemistry and consistency of that particular plant or soil. A plant known as reindeer lichen grows in Arctic regions. It forms a thick slow-growing mat of vegetation on forest floors and across Arctic tundra, providing a remarkably efficient surface for collecting elements from the atmosphere. Strontium 90 and cesium 137 were incorporated directly into the structure of this lichen.

The limited number of living species in the Arctic region offers very little variety in the diet of both animals and man. For nine months of the year reindeer live almost entirely on lichens. Many Lapps and Eskimos depend primarily on the reindeer as their food.

The result is that the radioactive elements are routed with remarkable efficiency into the bones and muscle tissue of the Arctic people. Long after atomic testing in the atmosphere had been banned, scientific measurements showed rising levels of cesium 137 in Lapps in Finland and Eskimos in Alaska. Thousands of miles from the original blasts and twelve months after the last tests, body deposits of cesium 137 had almost doubled in that one year's time.

The case of mercury contamination of fish is another example of biological concentration that took most scientists by surprise. The extreme toxicity of mercury has been known for over a thousand years. Two kinds of mercury poisoning can occur. One, caused by metallic mercury and inorganic compounds, damages the digestive tract and kidneys, but this damage is reversible and therefore not fatal. The other type of mercury poisoning is caused by organic mercury compounds. This poison attacks the brain and central nervous system and the damage is irreversible.

Mercury is used in many industries in the manufacture of drugs, cosmetics, paints, pesticides, felt hats, various electrical devices, chemicals, and plastics. The waste mercury from these operations has usually been disposed of in natural bodies of water. Since the waste product is in the form of metallic mercury or inorganic compounds and since these forms do not dissolve in water but fall to the bottom, it was generally believed that they would not be absorbed by any living organisms.

However, in the past decade reports of mercury poisoning began cropping up around the world and sampling of aquatic life showed high concentrations of toxic types of mercury compounds. After several years of intense investigation, Swedish scientists discovered the hitherto unsuspected fact that certain bacteria were able to convert metallic mercury into organic mercury compounds. Algae then absorbed the bacteria, fish ate the algae, and man ate the fish. Through this food chain the concentration of the poison was increased as much as ten thousand times.

The case of mercury poisoning illustrates the danger of making assumptions on the basis of our present scientific knowledge. Biochemists thought they knew how mercury reacted with living systems, but their knowledge was incomplete. In the meantime manufacturers continued to dump large quantities of mercury into lakes and rivers, and no experiments were conducted to test the validity of the assumption that the mercury remained inert at the bottom of the body of water. As a result of this miscalculation, enormous pools of cast-off mercury now lie under the waterways of the world and no one knows how they can be safely neutralized.

The phenomenon of biological concentration has been understood for a long time and has been demonstrated over and over—from the presence of radioactive zinc in oysters and clams to DDT in hen's eggs and mother's milk. In spite of this knowledge, technologists responsible for adding a damaging factor to the environment still seek to excuse their action on the assumption that these toxic factors will be uniformly dispersed and rapidly dissipated. In most cases there is no attempt to verify this very doubtful premise or to understand the complex biological reactions which may result from the broadcasting of the waste product.

Research—Too Little and Too Late

In 1966, when scientists became alarmed about the danger of mercury poisoning, the Swedish government canceled registration of alkyl-mercury compounds for agricultural use. The only official American response to the threat of mercury poisoning was the sampling by the Bureau of Sport Fisheries and Wildlife of a few pheasants and fish along the Atlantic seaboard. *Negative results from this sampling convinced the bureau to look no further.* A wider survey of fish and wildlife over a broad spectrum of habitats would undoubtedly have revealed the mercury pollution problem in this country four or five years before it was finally discovered. It would have spared many victims of mercury poisoning who subsequently suffered death or lifetime injury to the nervous system.

There is only one way tragic mistakes of this kind can be avoided:

careful, long-term scientific studies to monitor the side effects of technological processes. Hastily conducted and poorly controlled experiments can actually do more harm than good.

Studies of this kind would most logically and most efficiently be conducted by the industry developing the technology, since that industry is the only one possessing the facts concerning the process and knowing the possible variations and alternatives. Unfortunately, such a study is not compatible with the competitive profit orientation of business enterprises. The free enterprise system in this country allows industries to initiate new technologies, to withhold information from the public concerning the nature of the side effects produced by the technologies, and to escalate the use of the processes without any scientific appraisal of their safety. The economics of the industrial system puts maximum emphasis on cost reduction and efficiency. Long-term scientific studies cost money and take time. From the point of view of the industries, they cannot be economically justified, especially if they demonstrate that a new technological process which offers cost reductions causes unnecessary damage to the environment. The result is that the majority of industries do not adequately monitor the long-term effects of the processes they initiate.

It is a sad fact that in recent years big businesses have often been guilty of prostituting science in an attempt to hide or distort the truth. They have capitalized on the layman's belief that the statements of scientists and doctors can always be accepted as impartial and proven scientific fact. But scientists are human beings subject to personal bias like anyone else, and not always immune to personal profit.

Ever since 1962 the $8-billion-a-year tobacco industry has been pouring money into an attempt to discredit the scientific studies that have shown cigarettes to be a health hazard. At the congressional hearings on the cigarette-labeling bill, thirty-nine medical authorities and statisticians testified on behalf of the tobacco industry (and were presumably compensated as consultants). Only ten physicians testified in support of the Surgeon General's report. However, the evidence presented by the Surgeon General's report indicated overwhelmingly that smoking is related to the incidence of lung cancer and other diseases. The experts hired by the tobacco industry presented no real evidence vindicating cigarettes.

In order to protect itself against phony expert testimony, the public should learn to distinguish between solid scientific data and vague statements, containing no factual evidence, or half-truths, which distort the facts. The following statement is typical: "There is no substance in tobacco smoke that has been proved to cause cancer, heart disease, or emphysema." This declaration is literally true. No *single* substance has been proved to cause these diseases but the combination of the many toxic chemicals in cigarette smoke has been shown to cause them. The statement is a half-truth and deliberately misleading.

Similarly, following the publication of *Silent Spring* in 1962, spokesmen for the chemical companies quoted measurements of declining concentrations of pesticides in lake water and soil residues as evidence that the chemicals were rapidly disappearing. But since these statements took no account of the fact that concentrations were simultaneously building up in aquatic life and in earthworms and other soil organisms, the evidence presented was very misleading. The so-called dissipation was really a redistribution and concentration.

The investor-owned electric utilities spend a remarkably small percentage of their gross revenues on research and development. In 1970 this percentage was less than one-tenth as much as the average research commitment of American industry as a whole. These facts have been brought to public attention by several official sources, such as the Energy Policy Staff of the Office of Science and Technology. Federal Power Commission Chairman John Nassikas also criticized the electric, gas, and oil companies for the small percentage of revenue they spent on research.

As a result of this pressure, the amounts allocated to research and development have been inching slowly upwards, but the last available R&D figures for all the electric companies averaged less than one per cent of their gross operating revenues. And much of this small research budget is used for the development of cheaper methods of operation.

Checking for safety is usually done in response to public pressure and only in the most cursory manner.

A token experiment of this kind was reported in an article entitled "Medical Evaluation of Man Working in AC Electric Fields." The authors made the following statement:

> It is well known that corona discharges not only cause power losses, but also may produce ozone and radiation harmful to human beings. Therefore, the ground area of the East Lima Substation and areas under the HV lines outside of the station were surveyed on two occasions, once during cool weather (December 4, 1962), and the other during warm weather (July 28, 1965). The weather was clear on both occasions and members of the medical staff of The Johns Hopkins Hospital were present. . . . There was no evidence of corona on the lines or associated equipment, not even the slightest odor of ozone.

The most unscientific feature of this so-called experiment is the fact that measurements were made on only *two* occasions. The normal procedure in *any* scientific experiment is to collect dozens or even hundreds of readings before the data is considered scientifically significant. Possible sources of error are checked out and statistical averages computed. But in this experiment two readings were obtained, both on days when fair weather prevailed. The experimenters recognized that the effect they were looking for was caused by corona discharge. They also knew that corona discharge occurs principally in bad weather. Yet when they found no evidence of corona on the lines in fair weather, they took it to be significant that they found no evidence of ozone. This experiment is comparable to testing for sunburn by exposing the skin on two occasions—a warm evening in June and midnight in December. The results of this experiment would show no evidence of reddening of the skin. But they could hardly be considered serious scientific evidence disproving the phenomenon of sunburn.

Furthermore, testing for ozone on the basis of odor is too subjective a procedure to class as a scientific measurement. Sensitivity to the odor of ozone varies by a factor of two or more; some people can detect it in concentrations as low as .02 ppm and others do not detect it until concentrations of about .05 ppm have been reached. There are far more accurate indicators for ozone, ranging from the rate of cracking of stretched rubber bands to a variety of sophisticated instruments.

Basing their statements on experiments of this kind (at which two doctors from Johns Hopkins were present) power company officials say that electrochemical changes produced by corona discharge "are deemed insignificant by recognized medical authorities . . ." "The significance of the chemical processes can be judged," they say, "by the fact that there are today about 20,000 miles of transmission lines operating at 345kv and above in this country, all of which experience corona discharge. Yet never to our knowledge have any biological effects been reported."

Since no scientific attempt has been made to assess the long-term effects of extra-high-voltage lines—effects such as accelerated aging, decreased litter size, reduction in the rate of photosynthesis, genetic damage, and increase in the incidence of lung disease and cancer—we must assume that the statements by the power company refer to immediate biological damage. In other words, they have received no reports of people dying from electrostatic shock or having sudden seizures of illness that could be directly related to oxidant poisoning.

Similarly, hundreds of thousands of people in this country smoke cigarettes, and we do not very often see anyone drop dead when smoking one. Yet we know from careful scientific studies that cigarette smoking does have a long-term damaging effect on the human system, increasing the incidence of lung cancer and heart disease.

Millions of people in our cities and on our highways are exposed to high levels of carbon monoxide, nitrogen oxides, and other toxic chemicals from automobile exhaust. The effect is not severe enough to cause them to slump over their steering wheels in traffic. But this does not prove that no damage has been done. On the contrary, we are now quite sure that these chemicals inhaled over long periods of time increase the number of auto accidents and incidence of respiratory diseases.

The effects which scientists believe result from long-term exposure to electrochemical oxidants are too gradual and too subtle for the average citizen to connect with the presence of high-tension lines. Can you imagine a farmer calling up Commonwealth Edison and saying: "I just want to report that my wife and I are growing old before our time. The petunias in my front yard are dying and my best sow had only five pigs in her last litter"?

Although questions concerning the safety of extra-high-voltage transmission have been raised in the press and in legal proceedings, spokesmen for the power companies answer with vague statements such as "our engineers have checked this out and found the effects to be negligible."

As Rachel Carson pointed out, we urgently need an end to false assurances and the little sugar-coated pills of half-truths. The public has a right to full possession of the facts. We are being asked to assume the biological risks and to pay the hidden costs. In the case of high-voltage transmission lines, the electric companies have the ability to force these risks on all the property owners lying in their path.

9

Field Tests, Country Style

It was an unusually warm day in June 1970. The temperature registered a sweltering 92 degrees in the nearest cities of Jackson and Gallipolis, but on this high green hilltop on the outskirts of Sparksville, Ohio, a fresh breeze was blowing. It stirred the white embroidered curtains at the Strasbows' windows and brought in the fragrance of the wild roses in full bloom along the old fence row.

This spacious white frame house with its acre of shaded green lawn, its weeping willow tree, its vegetable patch, its two red barns and ninety acres of cropland belong to Clovis Strasbow and his wife Ada. The acquisition of this property was the realization of a lifetime ambition for Clovis. As a child growing up in Sparksville he had watched the construction of the big house on this high knoll commanding a pleasant prospect of woods and fields, and he had dreamed of owning the place himself. Twelve years ago the opportunity to purchase presented itself and the Strasbows had saved enough money to finance it.

"Everything we have we got the hard way, working by the hour," says Clovis. "When Ada and I got married I was so poor my socks had holes at both ends. Now all we got is tied up in this place here."

Clovis still works by the hour five days a week for a local construction company. He farms weekends and evenings. Ada keeps a few chickens and Clovis raises vegetables. "I reckon those vegetables cost more than if we bought 'em in the store," he observes, "but they taste a whole lot better. I like to grab a salt shaker and go down there to that garden patch and pick me a real ripe tomato warm from the sun—boy, that's livin'!"

Clovis Strasbow is a vigorous forty-year-old with well-muscled arms and a deeply suntanned face. His blue eyes and ready smile usually radiate confidence and good humor, but on this Saturday morning he looked troubled as he walked up the drive from the mailbox and came into the house carrying the day-old edition of the *Columbus Dispatch* in his hand.

The article that had caught Clovis' eye was the same that had caused such excitement in the general store in Laurel thirty miles away. Clovis was only mildly interested in the controversy about the right-of-way through the state forest. He was more concerned by the description of the electrochemical pollution produced by such lines.

Nine months had passed since a representative of the Ohio Power Company had knocked on the Strasbows' kitchen door. Ada was busy that afternoon putting up the last of the sweet corn for their home freezer. So Clovis answered the door and cordially invited the agent in, leading him through the dining room that was the center of their home activities. Here Ada had spread out a dress pattern for cutting on the big oak table. Here Clovis often worked at his accounts at an office desk. Here he also practiced the guitar, which he played with a local combo once a week.

They passed on into the front room and sat down on the sofa. The agent spread his papers out on the coffee table, mopped his brow, and hesitated a moment. Glancing around the room, he remarked on the large collection of family photographs that hung over the electronic organ. Beside the organ Clovis' favorite recliner was arranged so he could stretch out after a hard day's work and watch TV.

Finally the agent brought out several documents and broke the news that the Ohio Power Company intended to construct a transmission line across the Strasbow property.

"Well I got news for you, too," exclaimed Clovis. "You're not goin' to run one of them ugly things across my farm!"

"There's no use fighting this," the agent replied. "People have tried before—they always lose. The power company has won every case it ever took to court. You might as well sign here now and take the money because I promise you that line's going to go right there between your house and the barns."

But Clovis was firm. He would not sign until he saw a lawyer and learned his rights. However, he remained pleasant and cordial, even pressing the agent to stay and share their evening meal. "Ada has some beans and cornbread cooking," he said. "It's pretty hot food. You'll have to put gloves on to eat that!"

The agent, however, departed without any supper or signature and the Strasbows heard nothing more from the power company for nearly a year.

"Ada," called Clovis, "come look at this write-up in the *Dispatch*." They sat down at the big oak table and went over the article carefully together.

"I sure don't like the sound of this, Ada," Clovis said. "Seems like something we don't want in our back yard. But what I can't make out is why don't the power company *know* how much of this ozone or oxidant stuff an electric line like that makes? Why can't they just go out and measure it?"

As they were talking, it occurred to Ada that the son of one of their friends had recently set up an electrical laboratory in Ironton, a town fifty miles south on the Ohio River. She wondered if he would be able to advise them. Clovis thought he might; perhaps he would even be able to measure ozone concentrations near the big new high-voltage line that crossed the river just west of town.

So that afternoon Clovis and Ada drove to Ironton and engaged a scientific consultant to make field tests of ozone and induced electrostatic charge under the only operating 765-kv transmission line in the world. The results of these tests, which Clovis made freely available to any interested parties, showed higher levels of ozone under the line than in other locations in the countryside. Unfortunately, the recorded ozone levels could not be directly related to other air pollution measurements because the ozone detector had been designed and built by the consultant himself and, as Clovis expressed it, the consultant did not have a string of letters after his name.

These studies, however, were just the beginning of a continuing research project conducted by Clovis Strasbow on the effects of 765-kv transmission lines. One day he chartered an airplane and a pilot and, with his scientific consultant, he flew the already-existing 765-kv line from Louisa, Kentucky, to Piketon, Ohio, taking notes on all the properties it traversed. Later Clovis called on many of these landowners and obtained their stories firsthand. He became a one-man clearing house for complaints and for people seeking help and advice. Many of the landowners were relieved to find someone to turn to. "This line is causing us fear and annoyance but we don't know what to do about it," was the usual reaction.

All this activity did not go unnoticed by the Ohio Power Company. Eventually they sent representatives around again to the Strasbows' home. They were surprised when Clovis explained his concern about electrochemical pollution in terms they were not knowledgeable enough to handle. "How do you know so much about ozone?" one of them challenged. "You a chemist or something?"

"Hell, no; I'm just a country boy," answered Clovis. "Never had any fancy schooling. I went down here to the local high school and Ada, she never finished high school. But after we found out what this line was going to do to us, we been studyin' up on it." He pulled out his file drawer and showed them an impressive collection of reprints of scientific articles. "By the time I'm through with you," he promised, "*ozone* is going to be a household word!"

Clovis Strasbow's personal survey of eighteen homes along the right-of-way of the 765-kv line revealed a wide variety of complaints. Of these eighteen families, more than half had noticed a strange chemical smell, particularly in the evenings and early mornings. Three people complained of shortness of breath and irritation of the throat and nasal passages.

Twelve families reported receiving strong electric shocks under varying circumstances—from farm machinery, buildings, fences, and even clothes lines. One woman received a severe jolt when she removed a nylon scarf from her head while standing beneath the line. Children were shocked while playing barefoot in the grass. One man said he was unable to paint his aluminum-siding house because of the shocks received through the brushes. Although all the houses were elaborately grounded, several residents complained of shocks from the plumbing when they turned water on or off. Two women dreaded to go to the bathroom because of the shocks received when they sat on the toilet.

One landowner attempted to install a gutter on a barn about 200 feet from the right-of-way and was almost knocked off his ladder by the shock from the barn roof. The building had been grounded already, but the power company attempted to correct the problem by more careful grounding. When these efforts were not successful, the power company turned off the voltage on the entire line so that the gutter could be installed on this farm. Later, this same man attempted to paint his barn roof, but the electric shocks were so frightening that he was forced to give up the project. The power company suggested that he ground himself by running a chain down his pants leg and allowing it to trail behind him on the barn roof.

All the property owners along the right-of-way were bothered by the loud crackling and roaring noise of the line, loud enough to wake them from a sound sleep when the line was suddenly energized. In bad weather it sizzled and popped "like fritters fryin' on the front burner." The owners of a new one-story house located about 100 feet from the edge of the right-of-way were so disturbed by the noise that they called the power company several times at midnight and two in the morning. The company finally came around and installed "mufflers," little wire baffles every few feet between two of the conductors in each bundle. Hundreds of these mufflers were installed. The operation required two men working several days with a 100-foot crane. In spite of these mufflers, however, the line still makes enough noise to be a constant irritation to the family living beneath it.

At night in damp weather the line glows with flickering blue lights. Three people mentioned the fact that it draws lightning. Dazzling strokes occurred frequently near their homes and they lived in dread of summer storms.

All but one of the residents complained that their radio and TV reception was very poor. Two families had had the foresight to include in their agreement with the power company a guarantee of no degradation of radio and TV reception. For these families the power company had built a special receiving tower on a nearby hillside and had connected the tower to the TV sets with an underground cable. In spite of this special accommodation, however, on the day one of these families was visited the TV was inoperative. Some difficulty had developed with the underground cable. The power company had been called two days before but had not yet corrected the problem. Those who had not protected themselves with a special guarantee did not receive such preferred treatment. In answer to one complaint about poor television reception, the power company replied that the TV set must be old.

There were numerous reports of biological damage to people, animals, and vegetation under the line. A small grove of white pine trees showed poor growth and yellow needles. House plants and pear trees were reported to be dying. According to one landowner, horses running in his field under the line had all the hair and whiskers burned off their noses and several men working under the line had hair burned off their arms. One very interesting episode involved a man who owned riding horses and found that he was unable to use them after the line was energized. Both horses and riders received severe shocks when they passed under the line. The horses jumped and shied, and the riders always received a jolt when they dismounted near the line. The power company sent representatives to investigate the complaint. They experimented with making special insulated reins and leads, saddles and bridles for the horses. But even these extraordinary measures did not solve the problem. Horseback riding was given up on this farm as too unpleasant and hazardous an occupation.

It may perhaps be coincidental that one of the eighteen families contacted in this survey had a child dying of leukemia. The disease was discovered after the child had been living for a year and a half under the high-voltage line.

Some of the stories of these residents had not been officially reported to the power company, but many had; and the reports must have come to the attention of high-level executives, as evidenced by the extraordinary corrective measures taken in a number of cases. But when the 765-kv lines are really operating under normal load will they be turned off for a farmer who needs to replace the gutter on his barn and will mufflers be installed for every complaining property owner?

In spite of the complaints obtained from only a small sampling of residents along the line, the Ohio Power Company continues to make the assertion that no reports of ill effects have been received. All the reports of biological damage are brushed off as being mere figments of the imagination. Of course, some of the stories may be magnified by fear and some of the effects reported may be unrelated to the presence of the line. But it is wrong to make either of these assumptions without careful investigation, especially since the effects are occurring under a line that is admittedly experimental in nature. The phenomena described by the residents demonstrate the presence of a strong electric field under the line, not only on the right-of-way but several hundred feet on either side, and yet the long-term effects on people living and working most of their lives in this electric field have never been adequately researched. At the time these lines were built and energized, there had never been a study made and published in any scientific journal of the concentration of photochemical oxidants under such lines in varying weather conditions, nor any consideration given by the electric companies to the health hazards posed by these high-energy chemicals. The tests authorized and paid for by Clovis Strasbow came closer to a scientific evaluation of the problem than anything that had been published by American Electric Power Company.

10

Abuse of Discretion

The environmental impact of the generation and transmission of electricity has become an increasingly important factor in determining the quality of life throughout the United States. Power plants and transmission lines add destructive chemicals to the air, waste heat to air and water, and radioactive elements to air, water, and land. Because of the discretionary powers vested in the electric companies by the law of eminent domain, all the decisions which affect the extent and distribution of these pollutants are made by the utilities. The choice of site, the routing of lines, the type of fuel used, and the technology employed to convert the fuel to electricity—these decisions affect the quality of the environment and thereby the quality of life in large portions of the country. The areas affected may be hundreds of miles from the areas served with the additional power. Failure to conduct adequate pollution testing programs and to show a responsible consideration for the environmental consequences of its decisions is certainly an abuse of the discretionary powers granted to the electric power industry.

Typically, decisions on these important matters are based solely on economic considerations.

Planning for Power Production

The price and availability of fuel resources determines the choice of generating facilities. Coal is the cheapest fuel generally available and is, therefore, the preferred resource. The more expensive nuclear energy is used where suitable fossil fuel is scarce.

From an environmental standpoint, however, the matter is much more complex. Each type of plant presents its own particular pollution problems. The burning of fossil fuels adds between 5 and 6 billion tons of carbon in the form of carbon dioxide to the atmosphere every year. Fossil fuel combustion also emits sulphur dioxide in varying amounts depending on the sulphur content of the fuel used. In spite of efforts to regulate the emission of sulphur dioxide, it has been estimated that electric generating plants are producing nearly 17 million tons of this noxious gas each year. Smaller amounts of hydrocarbons (perhaps one million tons) and of nitrogen oxides (four million tons) pour out of the chimneys of electric generating plants. Fly-ash and other particulates are also emitted. The term "particulate" covers a great variety of very small bits of solid matter suspended in the air. These range from simple elements like iron or lead to complex organic substances. Many of the particulates are dangerous; some are known to be carcinogenic. Mercury has recently been identified in the smoke from the combustion of coal. In 1968 it was estimated that electric-power generation added about five and a half million tons of particulates to our atmosphere each year.

The generation of electricity by either fossil fuel or atomic energy is intrinsically inefficient; so for every unit of electrical energy produced, several units of waste energy are expelled to the environment at the plant site. Atomic plants are somewhat worse in this respect than fossil fuel plants. For every unit of electricity generated by an atomic plant, approximately three and a half units of energy are discarded as heat. Fossil fuel plants produce two and a half units of waste heat for every unit of electricity.

The easiest and cheapest way to remove this heat from the plant is to use lake or river water for cooling the condenser coils and then to flush this warmed water back into the natural body of water from which it came. The extent of the damage caused by this thermal pollution depends on the size of the body of water, the natural circulation and temperature conditions, the presence of other pollutants, and the total amount of waste heat poured into it. Lake Michigan, for instance, is a large and cold body of water, but it already carries a heavy load of pollution from other sources. Power plants at twenty-nine sites now use its water as a receptacle for their thermal waste and generating capacity is expected to double in the next decade. If power consumption continues to rise at the present rate, and there is no great increase in overall efficiency, by the year 2000 the heat from power generation would be enough to raise by twenty degrees Fahrenheit the temperature of the total volume of water which runs over the surface of the United States in a year. Long before that point is reached, thermal pollution will have caused the death of many types of aquatic life and even the death of the lakes and rivers themselves.

Each species of aquatic life has been adapted by evolutionary processes to the variations in water temperature that occur naturally throughout the year. In water, temperature changes occur very gradually and encompass a narrower range than the extremes usually encountered in the atmosphere. This relatively even environment is undoubtedly one of the reasons why evolution did not develop self-regulating temperature systems for aquatic forms of life. They are cold-blooded and therefore particularly susceptible to variations in temperature. An organism's body heat is an important factor in regulating the rate of vital processes such as metabolism, reproductive capacity, and growth, as well as longevity. Generally speaking, the metabolic rate doubles with each increase of 18° Fahrenheit. When the metabolism increases, the need for oxygen increases and the rates of respiration and heartbeat also rise. These changes put severe stress on the organism; and each species has its own characteristic level of tolerance. The water temperature which is lethal to different species of fish varies between 77 degrees F for cold-water fish and 97 degrees F for southern species.

Before lethal temperatures are reached more subtle changes occur which affect the equilibrium of the aquatic ecosystem. The optimal temperature for any individual species is about 10 to 15 degrees below its lethal temperature and the balance maintained among the many different species inhabiting the same body of water is altered as the temperature of the water varies. During the hottest summer months many of the lakes and estuaries in the United States reach 70 or 80 degrees F, making them unfavorable for some types of fish. Most species of algae, on the other hand, grow best in warmer waters. Peak summer temperatures, especially when occurring in waters "enriched" with organic matter from sewage and other effluents, cause the algae to proliferate, using up more than their share of the oxygen dissolved in the water; and, since fish need more oxygen in warm water (because of their increased metabolic rate), these factors combine to favor algae and plankton growth. The water becomes clogged with weeds and thick with algae. The bottom fills up with decaying organic matter and the useful life of the body of water is ended.

The conditions leading to thermal pollution of our lakes and rivers are most critical when heat from electric generating plants is added to waters already warmed by summer to a peak approaching the lethal temperatures of some of the native species. A very small increase at that point, even a few degrees, may tip the balance in favor of the aquatic plant life and toward early death of the body of water. Unfortunately, peak generating loads demanded by air conditioning in hot weather coincide with the most unfavorable natural water-temperature conditions.

A number of mechanical systems have been devised to alter the distribution of waste heat from power plants. The extra thermal energy can be put into the air by artificial cooling lakes or wet cooling towers that make use of the principle of evaporation. When liquid water is converted into water vapor, heat is absorbed. However, this evaporative cooling process adds water vapor as well as heat to the atmosphere. When the vapor condenses again into rain, fog, or dew the heat is delivered back to the earth's surface. Thus the waste heat is spread over a wider region in this process. It is redistributed, not destroyed.

Water vapor is not ordinarily classed as an air pollutant. But water droplets suspended in the atmosphere can react chemically with other pollutants. For instance, sulphur dioxide and water vapor are converted into corrosive sulphuric acid mist, and water vapor acts as a vehicle to carry the pollution deep into the lungs. Large amounts of water vapor added to the atmosphere also produce climatic changes.

It is obvious that clustering of power generating facilities aggravates the thermal pollution problems. Local changes in weather are much more likely to be serious when six generating plants eject their plumes of hot humid air within a small geographical radius or within the same temperature inversion basin. While one plant might dispel its heat into a body of water without noticeable effect, twenty or thirty plants dispelling their heat into the same body of water can cause irreversible damage.

Site-planning, therefore, is of great importance in minimizing the impact of power generation. Siting should be studied on a broad regional basis, taking into consideration the other pollution sources that already exist in the area or are projected for the near future. However, the policy of the power industry is to site new generating facilities in locations offering the greatest economy of operation, and where they will arouse the least public resistance. In most cases these conditions are satisfied best in a rural location, preferably very close to a source of fuel and a body of water that can be used for cooling. Since these sites are usually far removed from the place where the power will be used, many additional miles of transmission lines are being built and are planned for the future.

This policy may seem reasonable at first glance. It particularly appeals to the city dweller who will receive the power without further deterioration of his immediate environment. He assumes that the pollution produced in the rural areas will be "rapidly dissipated," and that in any case it will represent a negligible deterioration of "pure" country air and water.

But neither of these assumptions stands up to close scrutiny. Although pollution levels in rural areas have not been monitored as systematically as they have in large cities, the few studies that have been made have shown high levels of pollution, particularly sulphur dioxide and oxidants, in certain large rural areas in the United States. The Kanawha Valley in West Virginia, the Great Salt Lake basin in northwestern Utah, and the San Joaquin Valley in California are among the most unfortunate areas in the country as far as pollution is concerned.

The map following page 111, published by the Environmental Sciences Administration, shows the number of days that widespread High Air Pollution Potential was forecast in each locality (1960-70). It can be seen that countrywide patterns of air pollution are influenced by the siting of heavy industry and power plants, by prevailing wind patterns, and by local geography such as low-lying valleys, where temperature inversions occur most often.

Although information on nationwide pollution problems is available to the executives who plan the siting of new plants, this information is rarely considered. Large electric plants to produce power for Los Angeles are currently being built in the desert of southern Nevada where the air circulation is blocked between the coastal range and the Rocky Mountains. Black Mesa, at the four corners where Utah, Colorado, Arizona, and New Mexico meet, is the site of six enormous new coal-burning plants. The two already in operation are filling the natural basin between the Rockies and the Jemez Mountains with pollution like "a backed-up bathtub drain," and the area's winds slosh the smoke around all the way from Taos to Albuquerque.

More power for the Chicago-Gary and Detroit industrial complexes will be generated in huge new installations on the Ohio River, on the edge of one of the most heavily air polluted areas in the United States and located precisely where prevailing westerly and northwesterly winds will carry the sulphur dioxide and moisture-laden air from the cooling towers into the heart of this air pollution pattern (see map). Four new coal-burning plants are under construction in southeastern Ohio and West Virginia, and the Federal Power Commission reports that more facilities planned for this area will bring the total new generating capacity up to 17,000 megawatts of power by 1990. This is enough to provide for all the present average electrical needs of Chicago and Detroit. It is in addition to the present generating capacity in southeastern Ohio and West Virginia that is now providing sufficient power for local needs. Obviously, this particular region has been earmarked to serve as a giant generating complex for distant industrial centers.

In trying to sell decisions like this to the public, spokesmen for the power industry make much of the fact that the plants are located well outside city areas; and they have repeatedly urged that plants located in rural areas be exempt from standards for the emission of sulphur dioxide. But measurements of sulphur dioxide taken in West Virginia and Maryland show episodes of concentrations high enough to cause serious damage to vegetation and other living things. Total oxidant levels are also dangerously high. Christmas-tree farms in this mountainous rural area have suffered extensive loss of growth and quality in their trees since 1968, a year after the start-up of a large coal-fired power plant in their vicinity. A statewide survey of air pollution damage to vegetation in Pennsylvania in the summer of 1970 revealed damage to vegetables, fruit, lawns, flowers, and forest trees amounting to a direct loss of more than three and a half million dollars.

During 1970 and 1971, samples of rainfall throughout the northeastern United States from Maine to New York were chemically analyzed by Dr. Gene E. Likens from Cornell University, Dr. Herbert Bormann of Yale, and Dr. Noye M. Johnson of Dartmouth. These samples were found to contain a proportion of corrosive acids from ten to a hundred times greater than the proportion usually present in rain. The investigators believe that the acids are formed when rain passes through air polluted with sulphur dioxide and nitrogen oxides. This acid rainfall eats deep holes in marble and causes visible damage to metal and stone structures. Its implications for the delicate molecules of biological systems have not yet been assessed. The occurrence of acid rainfall has been known since the 1920's, but it had been assumed that this phenomenon occurred only around industrial centers. These investigators found that it occurred throughout the en-

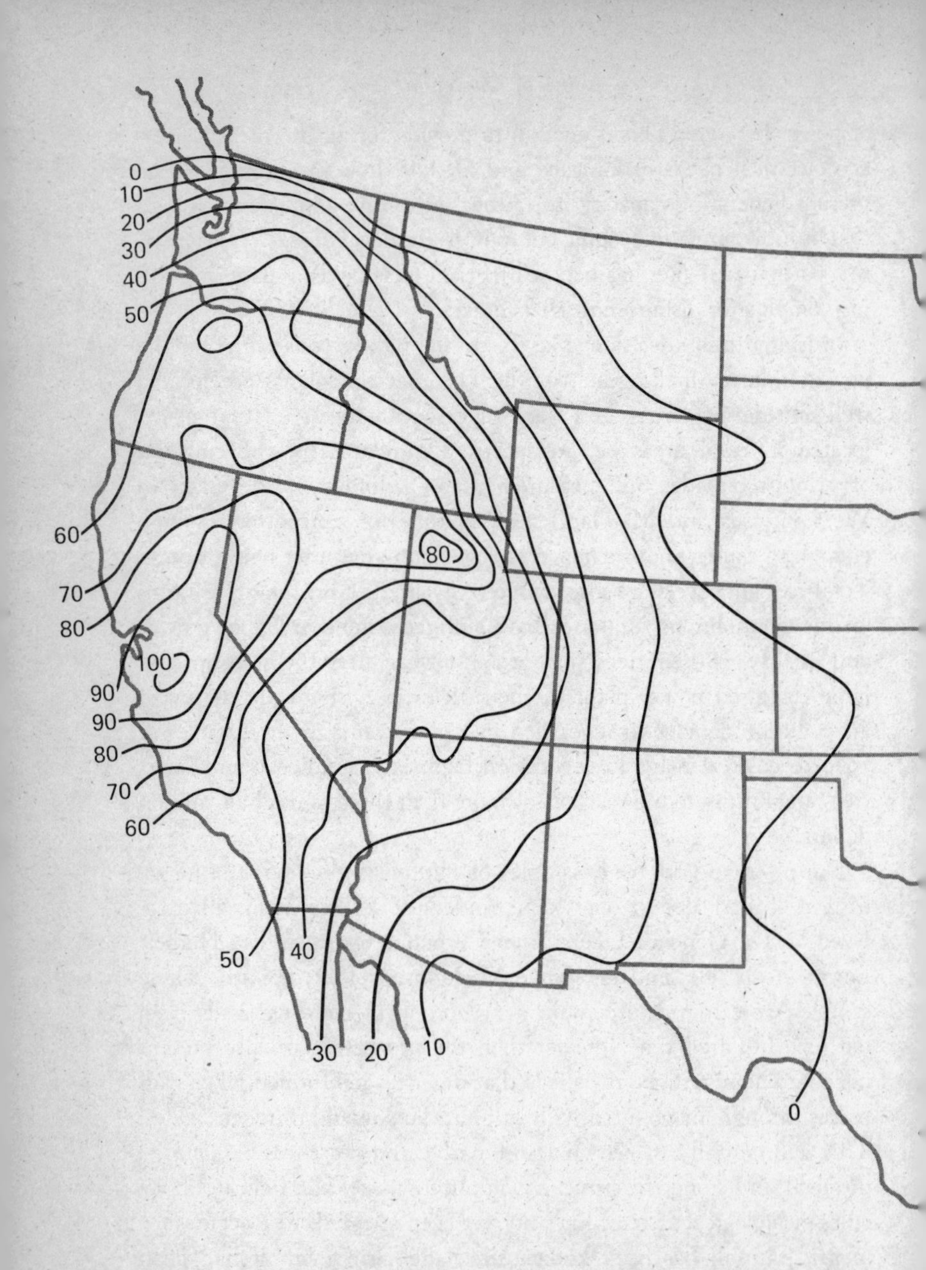

The number of days that widespread high air pollution potential was forecast are shown in this map from the Environmental Science Services Administration. The program began in the East on August 1, 1960 and in the West on October 1, 1963. Between those dates and April 3, 1970, there were 39 episodes in the West and 75 in the East. The numbers indicate the days a particular area was included in a HAPP (High Air

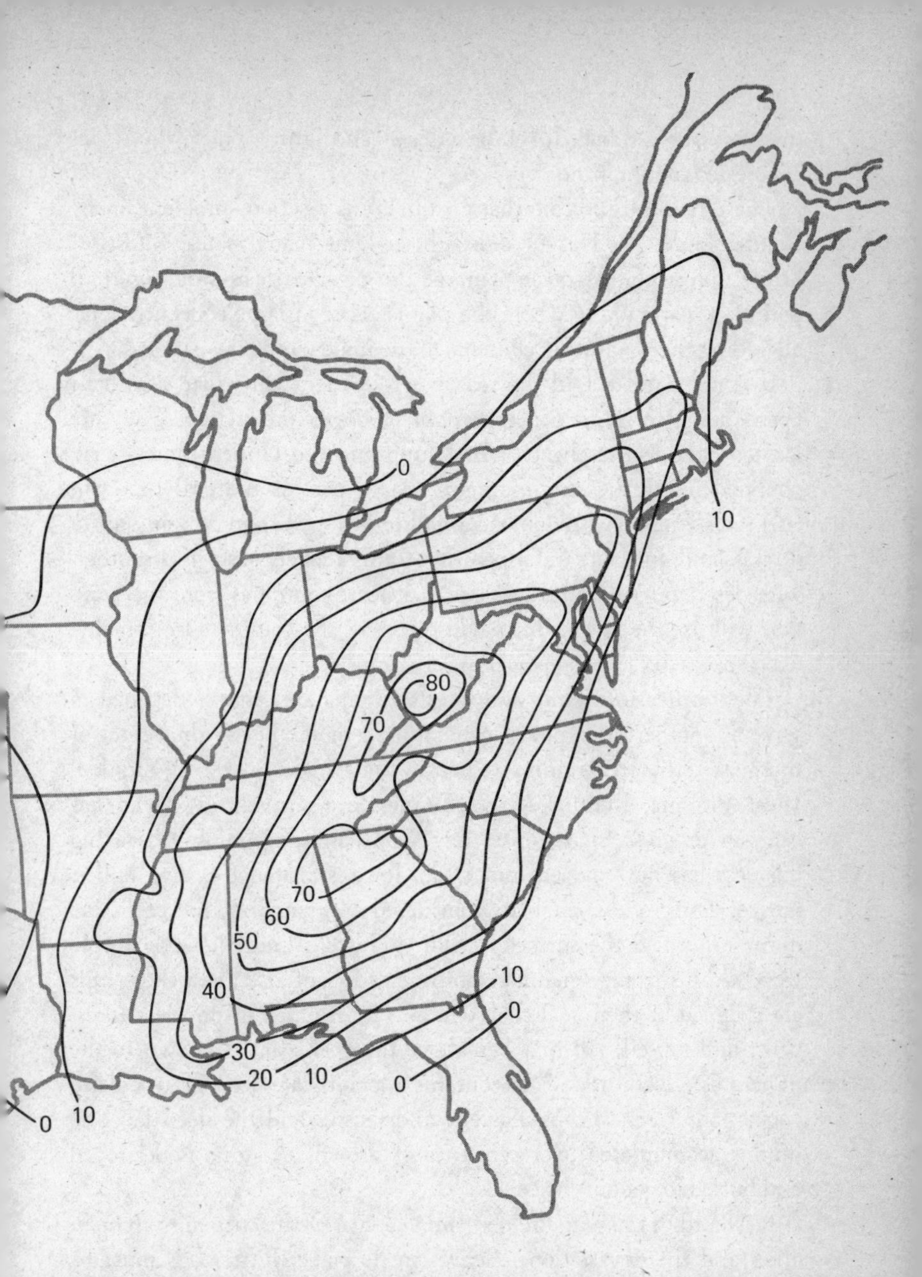

Pollution Potential) forecast. For example, between the line marked 0 and the line marked 10, the area was affected between 0 and 10 days. This map does not show the many additional days of bad air pollution weather of less than 75,000 square miles extent or less than 36 hours duration, or both.
(From Virginia Brodine, "Episode 104," *Environment,* Jan./Feb. 1971)

tire area they studied. It fell on villages and farms and on the slopes of forested mountainsides.

These facts all indicate that air quality is a serious problem in the countryside as well as in cities throughout many states. Siting of heavy industry and power plants is the principal cause of this rural pollution, particularly where the plants occur in large concentrations, all adding their polluting effluents to the same wind-flow patterns.

It is not hard to find the reason why certain particular sites have been chosen for large concentrations of power production. They are located close to the cheapest fuel supply in the United States—coal, obtained mostly by strip-mining. A large part of the coal that will fuel the six plants at Black Mesa will be stripped from a high plateau of grassland, juniper, and aspen that forms a lovely oasis in this otherwise desert region. Since this site is at least 500 miles from the cities that will use the power, lines will crisscross the country's last pristine open spaces to carry the power to "civilization."

The southeastern Ohio and West Virginia area that is destined to serve as another large power generation complex lies in the center of the most active strip-mining region in the United States. In Virginia, West Virginia, Kentucky, and Tennessee a strip-mining boom has been in progress for the past few years, and strip-mining as a big business has now moved into Ohio. It is estimated that five billion tons of low-grade coal, long considered too marginal for profitable mining, lie near the surface in Ohio. This coal underlies the deeply wooded hills from Cincinnati northeast to Canton. When these hills are stripped of their coal they will also be stripped of priceless forest cover and topsoil. All that lies above the coal seam is known to the miners as "overburden." It contains the intricate web of woodland vegetation: trees, shrubs, flowers, and mosses and the deep layer of humus accumulated over centuries of growth. All this is destroyed and buried to get out the coal.

It is hard to imagine the obscene violence of the strip-mining operation and the devastation it leaves in its wake. First roads must be slashed through the woods in order to bring in the massive equipment used in modern strip-mining—enormous augers, power shovels, bulldozers, and front-end loaders. The Giant Earth Movers (ironically nicknamed Gems) stand twelve stories high and have bodies so big that they even contain shower rooms for the crew. In one 24-hour shift a Gem uses more electricity than a town of 10,000 people. In one bite it can move 300 tons of earth.

These giant machines grind into the woodland on their massive treads. Their enormous blades press against tree trunks. With a rending sound the trees crash and tumble down the mountainside, together with the rich topsoil still cradled in the interlaced root structure torn from the earth. They are covered immediately by an avalanche of stones and yellow clay. Jack hammers scream as deep holes are sunk in the underlying slate and sandstone layers. The rock is blasted into fragments and the bulldozers shove them, too, down the mountainside. The cascade of rocks finally comes to rest on top of the "spoil banks" where lie the crushed forms of trees and mangled forest growth.

As cut follows cut, sheer cliffs known as "high walls" are created. These ugly jagged stone faces sometimes stand 90 feet tall. Below the high wall the bench of black coal that has been laid bare is loosened by more blasting. Then power shovels load the coal onto trucks, starting its journey to the power plants. Trains, often 100 cars long, shuttle the coal to the plants. In some locations moving belts carry a continuous ribbon of coal from the mines to the furnaces.

The strip-mining operation is the ultimate triumph of man's technology over nature. The superior efficiency of these enormous machines over human labor has made strip-mining a quicker and cheaper mining method than the underground mining that produced most coal until recently. New laws protecting the safety of the miner have increased the expense of deep-mining operations. Stripping produces three times as much coal per man as an underground mining operation and requires less capital investment. Such profit possibilities have created a veritable frenzy to cash in on the strip-mining bonanza. Ten years ago, about 120 million tons were produced by strip-mining every year; today at least 200 million tons are strip-mined annually, and the rate is escalating every month.

Many states have passed laws requiring reclamation of the stripped land, but with the exception of Pennsylvania's new regulation these laws are inadequate and poorly enforced. The multi-billion dollar coal industry is lobbying to block the passage of stricter laws in the other strip-mining states. Their tactics were recently described by Ohio's Governor Gilligan as "brazen and brutal" attempts to "blackmail" the General Assembly into weak legislation. Even when reasonable laws do exist, enforcement is very difficult. Inspectors can too often be persuaded to look the other way.

Most importantly, it simply is not possible to restore very precipitous country to anything like its original beauty and function in the ecology of nature. Land that is flat or gently rolling can, with conscientious reclamation techniques, be restored to usefulness. In order to do this properly, the topsoil must be skimmed off and preserved. After the coal has been removed, the pits must be filled and the rock and subsoil graded and compressed with compacting machines. Then the topsoil must be spread evenly on the surface and sprayed with fertilizer and seed. In England, Germany, and Czechoslovakia stripped areas have been treated in this manner and restored to fertility. The British have demonstrated that it can be done for a cost slightly more than one dollar per ton of coal mined. Strippers in this country, however, are using cheaper methods or none at all. It is estimated that two-thirds of our stripped mines have undergone no reclamation whatever. Our "best" routine efforts cost about 15 cents per ton of coal. Under these optimal practices a combination of grass seed, shredded straw, and fertilizer is sprayed over the raw subsoil. The precious topsoil lies buried in the spoil banks. In some cases pine seedlings are planted. The best results from these practices are gloomy treeless moors covered with spindly grass and crown vetch, or fledgling pine plantations that are a poor substitute for the hardwood forests they replace. "This stuff they're planting and calling 'reclamation,'" said an Ohio schoolteacher, "is like painting the face of a corpse."

Land that is steeply pitched cannot be successfully reclaimed. On slopes with more than 25 degrees of pitch the topsoil would wash away even if it were replaced, and grading is ineffective to prevent the land from sliding. The old established root system of the forest that held the land in place has been ripped out and the first heavy rain erodes the raw banks, washing quantities of earth down the bare mountainside. This silt and clay clog the stream beds, spoiling the natural drainage. Disruptions of soil layers cause many trace metals—copper, aluminum, iron, calcium, and manganese—to be washed out and concentrated in the strip-pits, man-made lakes left between the steep cliffs.

Mercury is a trace element found in coal beds and on river bottoms in strip-mined areas. Sulphur impurities in coal, when exposed to air, become converted into sulphuric acid, which is washed down into mine-pits and natural brooks. Throughout the strip-mined lands there are strangely colored waters—red, blue, green, and yellow. Yellow water indicates high sulphur content; blue is caused by aluminum, bright green by copper, red by iron. These chemicals are poisonous to all aquatic creatures; only bacteria can live in many of the sickly streams that flow through these devastated areas. Even a hundred years after the mining operator has finished, many of these streams will still run dead.

Likewise the soil that has been leached of its valuable nutrients will be useless for centuries. Its capacity to retain water has been seriously reduced. The hardwood forests that held the water and delivered it slowly to the land have been stripped away, exposing the whole watershed to extremes of flood and drought. When heavy rain falls on these violated hillsides, rushing torrents of silt and acid flow into the major tributaries and finally into the vast floodplains along the Ohio and Mississippi river valleys. The cities and farms along these rivers pay a heavy price for the profits of the strip-miners.

Undoubtedly there will be even heavier prices to pay in ways not yet anticipated. Nature has a habit of saving up our debts and suddenly presenting the long-overdue bill with heavy compound interest. It is a strange bit of poetic justice, for instance, that the only aquatic organisms that can survive in the heavily poisoned streams of the strip-mined watershed are bacteria—bacteria capable, perhaps, of turning the metallic mercury lying on the river beds into a lethal poison.

This poison, washed down to the larger rivers, may eventually find its way into the human food chain and turn this act of devastation back again at man.

Mining operators argue that the land they strip is worthless—its average market value is about $10 an acre. But actually this forested heart of our nation is priceless. Its elaborate web of root fibers holds the land and regulates the water flow; its layers of foliage help maintain a favorable balance between the oxygen and carbon dioxide in the atmosphere. And through the miracle of photosynthesis it creates a reserve of food and energy for the future.

Beyond these practical considerations the Appalachian Mountains in their natural forested state are beautiful. Within easy reach of some of our most populous cities, they offer many city dwellers an opportunity to escape from the friction of urban life to the green tranquility of the deep forest. They offer country dwellers an environment of unusual natural variety and splendor.

Belmont County, known as "Little Switzerland," used to be one of the most scenic regions in Ohio. More than half the acreage in this county has already been sold, leased, or optioned to coal strippers. "They are turning this beautiful place into a desert," lamented Representative Wayne L. Hays. His home is in a town that has been isolated on three sides by sheer, quarry-like cliffs where the strip-mine excavation stopped. Barns, silos, houses, and churches have been dynamited to lay bare the coal. Whole communities have been uprooted and the way of life that existed in these villages for generations has been ruthlessly destroyed.

"I would have expected this kind of thing in Russia but not here!" a telephone operator exclaimed.

The point of view of industry is represented by the remark of Ford Sampson, head of the Ohio Coal Association: "Are we to cut off electric power because some guy has a sentimental feeling about an acre of coal?"

The tiny village of Egypt was once the shopping center and meeting place for a wide rural community. No one lives there anymore.

There is an old Grange Hall still standing, an empty church, and a cemetery. Hanna Coal had the exterior of the church painted recently, but no one worships there anymore. The last family moved out of Egypt several years ago.

The same fate appears to be in store for the little town of Hendrysburg, which lies right on Interstate Route 70. Hanna Coal's giant "Gem of Egypt" has been digging night and day, practically in the backyards of Hendrysburg. Several times a day the town has been shaken by blasting and pelted with stones kicked up by the dynamiting. The shocks knocked pans off stoves and pictures off the walls. One man who made a hobby of building grandfather clocks wired his clocks to the wall. Some days as many as 80 shocks have been recorded. The plaster has cracked in the houses and the floors are sagging. When it rains water leaks through the cracked ceilings.

In five years the population of Hendrysburg has fallen from 800 to 500. Most of those who are left are middle-aged or elderly; many have already moved once from farmland that was bought by the coal companies. Some of the residents cannot afford to move again. Those who could move have no place to go. The farms and rural communities they are familiar with have been destroyed. There is no place left where they could re-create the way of life that they have lost.

Hanna Coal has an agreement with the State of Ohio and the Federal Highway Administration that the interstate highway will be closed to all traffic for up to twenty-four hours to permit Hanna to move its "Gem" across the four-lane road on a massive crushed rock and earthen dike when it is finished stripping the area around Hendrysburg and is ready to start stripping the farmland and wooded hills to the south.

The residents of the little communities on the south side are watching with growing apprehension the rapid progress of the mining on the north. They know that their turn is next and they are mobilizing to try to prevent the coal company from moving its Gem across the highway. A citizens' group in the pretty little town of Barnesville is attempting to protect their community from the fate that befell Egypt and Hendrysburg. They are seeking means to set up a green belt around the town, an area where mining could not occur and the

Ray M. Filloon for U. S. Forest Service, 1937

The land they strip is worthless.

Robert Charles Smith for *Environment*

countryside would be preserved. They fear that the strip-mining will foul their water supplies, damage their access roads, and leave them entirely isolated like a tiny raised oasis in the midst of a man-made desert. But the coal company has bought mineral rights and land rights up to the very edge of the town. Most of the green belt already belongs to the strippers.

While the destruction of the homes and farmland near Hendrysburg was taking place, American Electric Power Company applied to the Securities and Exchange Commission for permission to construct a 100-unit rental housing development in Cambridge, Ohio, just twenty-five miles west of Hendrysburg. "We feel," they said, "that utilities are in a unique position to help their communities meet their housing needs. We know our communities and we have both the resources and the willingness to undertake the job." The annual report that contained this statement included pictures of Martins Ferry, Ohio, "AEP System's newest community." Martins Ferry lies twenty miles to the east of Hendrysburg and Interstate 70 connects the two towns, bisecting Belmont County and Egypt Valley.

Looking at these facts it is easy to see that American Electric Power does have a unique understanding of the housing needs of this devastated region. Capitalizing on the necessity for new housing to replace the homes they have helped destroy, they plan to erect a multi-unit housing development. Built according to their formula for the American way of life, it will be another all-electric homogenized suburbia. Life in a planned community like this will be as different in flavor from the life in a real country community as a hothouse tomato is from a tomato grown in Clovis Strasbow's garden. But American Electric Power will make a profit from the project, which they will finance out of profits acquired from the sale of power generated from the coal blasted from the backyards of Hendrysburg and little Egypt and Barnesville.

Interstate 70 is also crossed by one of AEP's new 765-kv power lines. The line originates at the enormous Kammer-Mitchell power plants on the Ohio River, plants fueled by the coal stripped from this area. The line passes just to the north of Cambridge and heads diagonally across the farmlands of Ohio to connect with the lines leading to the Chicago and Detroit areas.

Concern for the Total Environment

Electricity created by this devastation of the Ohio countryside is already being marketed in Chicago. While the monstrous bulldozers work around the clock tearing at the hillsides along the Muskingum and Ohio rivers, a smooth voice comes over the Chicago radio several times a week: "People want to go where there are no chimneys, where there is no dirt to seep in and collect on windowsills. The next time you remodel install clean electric heating. Electricity creates no dirt and no fumes because there is no combustion. *Commonwealth Edison, concern for your total environment.*"

What is our total environment, Commonwealth Edison? Is it the four walls wired for electricity that produce the little cell of space that we call home? Is it the great heap of concrete and humanity that we call our city? Is it America, that place we used to sing about as beautiful with spacious skies and amber fields of grain? Or is it the whole earth, this planet so intricately and dynamically integrated that it is almost like a single living organism? Injuries inflicted in one region cannot help but affect the health of the whole earth. The deep wounds that make the streams of Kentucky and West Virginia run red will eventually affect the total environment of the people who live on the shores of Lake Michigan as well as those who live beside the cold waters of the Baltic and on the banks of the Ganges.

The "clean" electric power that the dulcet voice persuades us to plug into does create dirt and fumes—a few hundred miles away. This power is responsible for the crushed forms of the sycamore trees and rhododendron bushes that lie rotting in the spoil banks along the dead rivers. It causes the plumes of dust that rise from the stripped fields and drift eastward to settle on the sidewalks of Wheeling and the windowsills of Morgantown. It causes the fumes laden with sulphur dioxide, nitrogen oxides, and water vapor that hang like a heavy smothering blanket over the valleys of West Virginia and Tennessee.

The public utilities are making cheap electric power available to some parts of the country at the expense of others. They are deceiving the public with misleading statements about "clean" electric power, and separating cause from effect so that the people who are plugging in color TVs cannot see the price that someone else is being forced to pay for their power.

Little Bill

Many people who lived through the 1950's in the Midwest can vividly remember the pert little bird that flashed onto their television screens at frequent intervals, hopping in on the heels of Loretta Young and interrupting "Victory at Sea." A catchy tune accompanied his appearance and carried this message: "Electricity costs less today, you know, than it did twenty-five years ago. A little birdie told me so —Little Bill!"

Little Bill was a delightful ad, and furthermore, his message was true—in a limited sense. Electricity did cost less in 1950 than it had in 1925; and in 1972, in spite of about 50 per cent general inflation, electricity costs only 10 per cent more than it did in 1950. A remarkable achievement, one might say, and indeed it would be if the rate we were paying represented the true price of the energy we are buying.

In supplying low-priced power the utility companies have used the same mass-production principles that Henry Ford pioneered in the manufacture of automobiles. By producing an inexpensive commodity they can sell more of it; and by making more of it, they can use their equipment more efficiently, reduce the price even more, and so on. A full-page advertisement run by the power companies in the spring of 1970 explained the method:

> The nation's Investor-Owned Electric Light and Power Companies spent billions for new equipment which produces and delivers electricity more economically. And the fact that you are using more electricity than ever before enables us to use these facilities more efficiently. In meeting the nation's increasing needs for electricity within the free enterprise system, our concern is not just to give you the most reliable service possible. We want to go right on serving you at a price that lags far behind the cost of living. That's good business.

The advertisement pictures children playing on a hillside with ankle-deep grass and lovely trees outlined against a clear sky. What does this scene have to do with the production of electricity? Are these some of the trees that will be cut down to make way for transmission towers 130 feet tall, carrying voltages so high that the children can never again fly kites or run barefoot on this grass? Or is this one of the hillsides that will be stripped next week to tear out the coal to feed the furnaces that pour out fumes and darken that clear sky? Is this one of the hillsides that will be left bare of vegetation and scarred with erosion because reclaiming it would cost too much?

R. W. Hatch, president of the Hanna Coal Company, explained the rationale of this economic policy as follows: "We [the strip-miners] oppose unreasonable restrictions that, if imposed upon us, would have increased our costs to such an extent that we would have priced ourselves out of the market place and deprived the public of the low-cost power it sorely needs."

This policy is the real basis for cheap electric power. Every step of the technology, from removal of the fuel from the ground to delivery of power to the consumer over marginally designed transmission lines is performed in the cheapest possible manner. Every step involves environmental degradation which could be avoided or corrected, with the cost included in the price of the product. If it were, as Mr. Hatch so clearly sees, the product might not be competitive with other, more honestly priced commodities. This deceptive pricing policy makes it possible to sell electricity so cheaply that it appears to be a bargain. The cheapness escalates the demand and undercuts all competition. The result is a rapidly rising rate of consumption and a rapidly accumulating environmental cost which is every day collecting compound interest.

This is good business, the power companies' ad tells us. Yes, indeed it is good business—for the power companies and their suppliers. During 1969 and 1970, when most businesses were experiencing a reduction in profits or even operating at a loss, American Electric Power reported steadily rising profits. In 1969 its net profits of $106.3 million were 6.5 per cent higher than in 1968. In 1970 its profits rose to

$116.9 million, a 10 per cent increase. In 1971 the increase was 16 per cent, to a net profit of $134.9 million. The electric power companies are riding high on the wave of a rapidly growing demand for electricity, a demand which they themselves have created by their pricing and production policies.

Cheaper by the Dozen

The rates charged for each kilowatt-hour of electricity are graduated almost as steeply as the income tax rates, but in the opposite direction: they favor the wealthy at the expense of the poor. Most heavy consumers pay less than a third as much for each kilowatt-hour as the small user does. For example, in 1972 Commonwealth Edison's residential rates start at 3.65 cents per kwh after an initial service charge. When quantities over 450 kwh are used in a month, the rate drops to 2.05 cents. But the person who heats his home with electricity pays only 1.10 cents per kwh for all quantities over 325 kwh. This rate applies not only to electricity used for heating but to all the electricity used in this home: the power that runs television sets, electric can-openers, and filters for the swimming pool. The power for air conditioning is also billed at this bargain rate. Therefore, the owner of a large home with electric space heating can enjoy the luxuries of his "all-electric living" at a much cheaper rate than the small consumer pays. Two or three rooms can be air conditioned for this favored customer at the same price the average homeowner pays for cooling one.

A study of the rate schedules makes it apparent that the low price for massive consumption is a promotional device to stimulate higher regular use and to undersell competitive energy sources. In applications where electricity is not economically competitive with other methods of doing the same job, electric rates have been artificially reduced until the prices of the two methods are comparable. Electric space heating, for instance, requires the consumption of approximately twice as much fossil fuel as would be required if gas or oil were used to fuel a well-designed home furnace. By marking down the charge for electric space heating, the electric companies have made the two methods roughly equal on a total cost basis. This practice amounts to a subsidy for electric heating, a subsidy paid for by the average electricity-buying American citizen and by every American citizen in terms of unnecessary degradation of the environment. If electric heating had to compete honestly with other forms of heating it would immediately be apparent that it is an extravagantly wasteful way to warm our homes.

However, electric space heating is being extensively promoted by the power companies. Through full-page advertisements and radio and television commercials the public is constantly bombarded with seductive statements about the advantages of the electric climate. The power industry has defended its promotion of electric heating by asserting that winter use of power balances summer use of air conditioning so that generating equipment can be used more uniformly throughout the year. However, the rate system is so designed that the use of electric heating actually encourages more electric air conditioning. People who heat with electricity can cool with electricity much more cheaply than those who heat with gas or oil. Under these conditions, the promotion of more electric space heating does not serve to equalize the summer-winter load but raises the level both summer and winter. Furthermore, the promotion of electric heating in the face of the pollution problems created by power production is a particularly flagrant example of what appears to be a deliberate campaign to mislead the public.

"Buildings with *the electric climate,*" declares one of these ads, "use the cleanest source of energy there is. *Flameless* electricity. There's no combustion! Therefore, buildings with *the electric climate* put nothing in the air around them!"

That statement is a misrepresentation of the facts. More than 80 per cent of the electricity in the United States is made by combustion of fossil fuels. According to an estimate made by the Scientists' Institute for Public Information, the combustion that generates "the electric climate" leads to three times as much carbon dioxide production, a several-fold increase in the oxides of nitrogen, and a several-fold increase in sulphur oxide releases to the environment over those which would result were natural-gas space heating employed. One all-electric building can have a power demand greater than that of a city with a population of 60,000 people. "It seems reasonable," says one commentator, "to ask whether electric space heating should be permitted at all, to say nothing of being touted as being 'clean.' "

Disposable Power

Not only residential rates but also those charged to business and industry offer bargain prices for big consumers. Two kinds of charges are made—one based on the maximum power drawn for any thirty-minute period during the month and the other based on the total power used. Both of these charges are graduated, giving a price advantage to the large consumer. Typical bulk rates in 1972 are 0.7 cents per kwh (Ohio Power), 0.6 cents (TVA), and only 0.2 cents for the hydroelectric power of the Pacific Northwest. These rates favor the competitive position of industries that use extraordinary amounts of power.

The primary metals industry is one of the largest consumers of electricity in this country. One steel mill can use as much electricity as a city of 200,000 people. But the manufacture of aluminum is the biggest consumer of all, accounting for about 10 per cent of all industrial power use. To make one ton of aluminum requires about 17,000 kwh of electricity. In comparison it has been estimated that the energy equivalent required to make one ton of steel is about 2,700 kwh. Obviously, aluminum production is an expensive process in terms of the energy required, and it consumes a disproportionate share of our total energy resources.

However, the low rates for electricity and the graduated price structure obscure this basic economic fact. The manufacturers of aluminum buy enormous quantities of power very cheaply; therefore, the product can be sold at a bargain price. Aluminum is now so cheap that we can afford to throw it away. We can make beer cans out of it and disposable pie tins and wrappings for hundreds of different products. The energy that goes into the refining of aluminum is paid for at a cut-rate price, but the cost to society of the pollution caused by each one of those kilowatt hours is not billed at a bargain rate.

Given the cost advantage of cheap electric power, the manufacture of aluminum is one of the most rapidly growing industries in this country. Production increased 453 per cent between 1946 and 1968. The spectacular growth of this one industry is illustrative of the factors pushing up the demand for electricity to the present exponential rates of increase.

New technologies for the making of steel threaten to escalate the use of power in producing this metal also. The basic-oxygen process consumes nearly 5 times as much electricity as the open-hearth method, and the electric furnace requires 65 times as much energy. (These figures were quoted with satisfaction in the report of the annual meeting of stockholders of Commonwealth Edison, April 2, 1971.) The reason for going to these high-electricity-consumption processes is that the open-hearth method creates serious air pollution problems. But in estimating the environmental advantages of the alternate technologies, it is essential to include the pollution caused by the generation of the additional power they demand. On this basis it would be difficult to justify the electric furnace that uses 65 times as much power as the open-hearth method and 13 times as much as the basic-oxygen. Financially, the electric furnace is brought within competitive range of the other methods by cheap bulk rates for the massive users of power.

This price structure, which encourages more and more extravagant use of electricity, cannot be justified on the basis that the 0.7-cent kwh costs the company only a fifth as much to produce and deliver as the 3.65-cent kwh. Large blocks of power cost the company a little less per kilowatt because of the fixed expenses of reading meters and sending out bills, but these small fixed expenses are largely covered by the minimum service charge and do not justify any such drastic reduction in price. On the whole, changes made in the rate structure over the past forty years have been in the direction of increasing the differential between the rates for the small and the large consumer. The result is that the large consumer is encouraged to use more power.

Large successful companies like American Electric Power and Commonwealth Edison are certainly well aware of the simple economic principles involved here. By reorganizing their rate structure they could accomplish a great deal toward damping the overall rate of growth, as well as equalizing the seasonal and diurnal load patterns. Residential rates could provide a basic standard-of-living ration for each household at a reasonable rate, with escalating prices for extraordinary uses over and above this ration. Such a rate structure would provide an incentive for the private citizen to conserve power.

Industrial and commercial rates could favor the use of power during hours of normally lower demand; industry would find ways of taking advantage of this saving. These methods are so elementary and readily applicable that it is impossible to avoid the conclusion that they are not employed because they would not serve the interests of the power industry. The only logical conclusion to be drawn from the present rate structure is that it is planned to promote rapid growth in general consumption and especially to favor certain prodigal uses of power which offer maximum growth potential far into the future.

Electric heating, for instance, offers the largest single market for more electric sales in the next decade. The goal of the utility industry, as reported in a Federal Power Commission Survey, is 19 million electrically heated homes by 1980. The power required by these homes would be equal to the entire residential use of electricity in 1960. But at the same time it is pushing electric space heating by hard-sell advertising and subsidy prices, the utility industry is protesting that it can hardly match the rapidly growing demand for power. The people *need* more power, the industry spokesmen declare. They *demand* it. Electricity consumption will double in the seventies.

Any advertising man is well aware of the psychological power of positive assertion. Tell people they need something, talk up its advantages in glowing terms, make it available at bargain price, and—presto—you have created a demand. The assertion of the power companies that "demand" for electricity will double in the 70's is a self-fulfilling prophecy. Since the companies have already ordered the plants that will produce this extra power, they now have a large financial stake in seeing that the prophecy becomes a fact. But the power industry does not seem to have given any responsible thought to where these policies are leading us.

Our All-Electric Future

We are now using in this country five times as much power as we were using in 1950 (four times as much per capita). Has our standard of living improved comparably? I am not sure that it has improved at all. We have more disposable aluminum items, more air conditioning and color television. But degenerative diseases such as cancer and emphysema are on the increase. Our cities are becoming less habitable and our land less beautiful.

By 1980 we will be using ten times as much power as we were in 1950; and if this same geometric rate continues, by the year 2000 we will be using 40 times as much electricity as we were using in 1950. (The electric industry's own estimate is somewhat more conservative: 25 times the 1950 consumption.) In the meantime, what will have happened to our standard of living?

A great deal of the increased power production will go into aluminum and steel items designed either for immediate disposal or for rapid obsolescence. Some families will have added work-saving devices such as washing machines and dishwashers. And, of course, more people will have that "flameless electric heat." By the time the majority of people have electric heating, the price subsidy will have virtually disappeared. Little Bill will have grown fat over the years and the cost of electric heating will have become apparent. But by that time most people will be living in homes without furnaces and will be committed to the "unpolluting" electric climate. To power this electric living we would need at least eight times as many electric generating plants as we have today, and this factor alone (assuming that other polluting sources remained constant) would increase by several times the present air pollution levels.

The life of an electric generating plant is approximately thirty years; there is a five-year interval from the time it is ordered until it is operating. The plants on order today will still be serving us in the year 2000. Since these plants have already been planned, they will be built to the same pollution-producing design that is causing many of our present environmental problems. Therefore, it is extremely unlikely that the pollution caused by the production of each kilowatt of electricity will change much in the next twenty or thirty years.

The rapid deterioration of air quality during this period might be delayed by outlawing automobiles using the internal combustion engine. Transportation in cities could be solved by providing rapid transit service and cars that run on batteries. But that would require more electricity and more power plants.

The waste disposal problem will also have become more critical (all those aluminum pie tins and junk yards of old automobiles). But these problems could be relieved by giant machines to compact and shred the waste. Of course, these machines would be electrically operated; and that would require more power plants.

The earth's climate will very likely be hotter and more humid (from the waste heat and water vapor created by the power plants). But that won't be any problem because we will have air-conditioning, which we could run for more of the year; and that would take more electricity.

Since the air pollution everywhere will probably exceed maximum permissible levels, it will be unsafe to spend any time out of doors, and all our living spaces—our homes, factories, automobiles, trains, perhaps even the sidewalks—will have to become part of the total electric climate. The air will have to be filtered and conditioned in order to be safe to breathe. That will take more electricity and more power plants.

The present course of action, far from leading us down the highway to progress, leads to an accelerating treadmill where we have to run faster and faster just to stay even. Somewhere along this road, the tolerance level of the public will be exceeded. A halt will be called and priorities established for the use of power.

In a sensible priority system electricity would never be used to correct other pollution problems unless it could be shown that the total degradation of the environment, including that caused by the generation of the additional power, would be reduced by this method. The electric industry, for instance, is now burning a great deal of natural gas in an attempt to meet air quality standards. (In 1971 one-fourth of the electricity made in this country was generated by natural gas.) But natural gas is most efficiently used to heat homes and water heaters directly. It is a relatively clean fuel and is in very short supply. Using this desirable fuel to generate electricity reduces its energy value by at least one-half and causes twice as much total air pollution for energy delivered. An intelligent energy policy would prohibit the use of natural gas for the generation of electricity, thus making gas available in more adequate quantities for home heating. This allocation of energy resources would reduce the need for increased electric generating facilities.

Another way in which the demand for electric power might be altered is through a change in the rate schedule. These rates are reviewed regularly by the public utility commissions. Public hearings are held and citizens' opinions can be aired. Citizens have a right, for instance, to protest against the discriminatory practice of providing cut-rate power to electrically heated buildings. A change in this factor alone would go a long way toward curbing the runaway demand which the power companies protest they are desperately trying to match.

New York City's Environmental Protection Administration recently released a study recommending ways to curtail the use of electricity, such as: increased rates for large users, a surcharge on use of electricity during hours of peak demand, and an electrical-use tax to reflect the environmental costs of power production. Executives of the power industry, however, are less than happy about such suggestions. "It has been suggested," said Donald Cook, president of American Electric Power, "that environmental considerations may require a reduction or stabilization in the demand for electric power. This is sheer nonsense."

John W. Simpson, president of the Power Systems Company of Westinghouse Electric Corporation, described his objections in more detail:

> Suggestions that we weight our rate structures to penalize large users of electric power, notions that we freeze levels of power consumption, recommendations that we begin to cut back on our use of electricity, all of these things, well-intentioned though *some* of them may be, are absolutely *wrong* and dangerous. To pull in the reins of our economic growth, to push down our living standards, to paralyze our society, would be a sure guarantee of national disaster and the swift demise of the United States as a nation of any consequence whatever.

It is not surprising that the industry resists any attempts to restrain the accelerating growth of demand for electric power. The plans, financial policies, and long-term commitments of these companies have already been made, predicated on the assumption of a doubled demand for electricity by the end of this decade. Bonds have been floated, million-dollar orders placed; engineers and boiler-makers are hard at work constructing the facilities that will tap this demand. Obviously, the companies will resist any fundamental change which would alter the level of power demand in 1980.

For the next year or two, however, there may be a tendency for the industry to go along with the policy of softer sell. Delays have occurred in bringing the new facilities in on schedule. Plants that should have been in operation in 1972 and 1973 will not be ready until 1974 or 1975. Several large utilities will not have the capacity they had planned to have during the interim. For this reason, they may appear to acquiesce in the public demand for less promotion of power use. Softening the hard sell will help tide them over the transition period.

Consolidated Edison, for instance, the hardest-pressed of all the large utilities, announced that it would cease all advertising in 1971 and, in fact, would turn to a policy of anti-sell. But it is safe to assume that, as soon as its new facilities have been completed and the boilers are building up steam, anti-sell policies will be discarded. Promotion campaigns will again be undertaken until the new plants are producing at full capacity and "the public demand must be met" by building more. Anything as fundamental as a change in the rate structure is likely to encounter enormous resistance from the industry, because that cannot be turned on and off as easily as an advertising campaign.

Most of the delays have been caused by the installation of inadequately tested equipment. In a desperate attempt to take advantage of the accelerating demand they themselves created, the power companies have been rushing into service scaled-up models of older generating and transmitting systems. They have gone from the 200-megawatt generators in the mid-1950's to 1,200 and 1,300 megawatts today, without normal testing for reliability of operation. "Now we are doing our testing in the utility plants," a White House expert is quoted as saying, "instead of in the manufacturing plants." The result has been frequent breakdowns of large generators such as Consolidated Edison's million-kilowatt "Big Allis" in New York City.

The extra-high-voltage transmission lines are another example of inadequately tested equipment put into service before their effects are understood. And the public is required to act as guinea pigs. (Call us up and report if there is any biological damage!)

The power companies resent citizen groups who question the safety and environmental impact of their projected installations. They blame the conservationists for the threatened power shortages that are the result of delays in construction of new facilities. But most of these delays have not been caused by citizen groups. Several official surveys have been made of the causes for delays in bringing large generating plants into service. Of 55 plants that were behind schedule between 1966 and 1968, only four delays were found to have any relation to environmental objections. Another more recent survey showed that between 1966 and 1970 only 7 per cent of the delays were caused by conservationists.

Many reputable authorities in the energy field agree that the principal reason for power shortages is mechanical failure due to technological errors. Former Federal Power Commissioner Carl Bagge charged at the April 1971 American Power Conference that the problem "was engendered by a monstrous sense of intellectual and technological arrogance which ignored not only the limitations of technology but even more importantly, the limitation of the vision of its high priests." Environmentalists are used as convenient scapegoats to cover up the mistakes of the power industry executives.

In the small percentage of cases where delays have been caused by groups of concerned citizens, their questions could have been answered without expensive delays if the power companies had done a conscientious job of studying all the safety factors and informing the public about these studies when building plans were announced. In nuclear plants the reliability of the system is of crucial importance for the protection of the public. It is the citizens of nearby cities and surrounding countryside who are being asked to assume the risks. If the companies had won the public's confidence by using the best available techniques to protect both the people and the environment, rather than waiting until such action was forced upon them, they would not be experiencing this growing resistance to their plans. The people have learned that unless they question the various safety factors, unless they demand public hearings and disclosure of the well-guarded secret plans of the power companies, marginal designs and untested technology will be forced upon them.

Behind the policies of the electric companies lies the tacit assumption that some risk is permissible in order to speed the growth of industrialization. They have the power to decide how much risk is permissible and who will bear it. By using this power to perpetrate an unequal distribution of the costs and benefits, by using it to install equipment that maximizes economy rather than safety, and finally, by sweeping us all into a maelstrom of wasteful consumption, the utility companies are grossly abusing the discretionary powers vested in them by the American people.

11

Ohio Power Company versus Clovis Strasbow

On the 23rd of January, 1971, Clovis Strasbow's family was served with a summons to appear in the Court of Common Pleas of Vinton County on a complaint filed by the Ohio Power Company in its attempt to appropriate a right-of-way for the transmission line across their property.

A number of times over the past year, Clovis had consulted with a lawyer in the nearby town of Wellston but had not received much encouragement for his primary objective of preventing the transmission line from crossing his property. The lawyer pointed out that there was no precedent for a decision of this kind in condemnation proceedings. He thought the best they could do would be to demand payment for the entire Strasbow property, maintaining that the value of the whole place would be completely destroyed by the erection of the line. He also suggested that Clovis obtain an appraisal of the property.

The appraiser hired by the Strasbows recognized that this piece of property was underlaid by several veins of coal which added considerably to its worth. Four veins of coal are known to lie under this portion of Vinton County, and No. 2 vein is especially valuable since it is very low in sulphur content. With the current demand for low-sulphur fuel, the coal from this vein would bring premium prices. Various companies, among them the Ohio Power Company itself, were at that very moment working hard in Vinton and Meigs counties to buy up land and mineral rights. Several of the Strasbows' neighbors had been approached, and the prices offered were being rapidly escalated to win over reluctant property owners.

"You might just as well sell out," one of the neighbors had been told. "You won't want to live here anyway. The whole county will be covered with soot and coal dust." Earth-shaking blasts of dynamite and the crash and rumble of giant strip-mining machines like Ohio Power Company's Big Muskie would make life intolerable for many miles around.

Residents of Vinton County did not have to be told these facts. They had already had a taste of strip-mining. They knew that when the miners and machines went away they would leave a scene of devastation resembling the mountains of the moon. Thousands of acres within a 10-mile radius of the Strasbows' property had been stripped; and, although most of this mining had occurred more than fifteen years ago, the land was still barren with precipitous mud cliffs and ugly pools tainted with red and yellow acid.

Raccoon Creek, which had run clear and sparkling when Clovis fished there as a child, is now dark muddy yellow and so polluted with sulphur that no fish live there. In heavy rains the water pours down off a number of bare gullied hillsides that drain into Raccoon Creek and cause this sizable tributary to flood regularly. Several valleys near the river have been turned into acid bogs where the gloomy skeletons of hundreds of dead trees still stand. Residents of these counties are within easy driving distance of Belmont County, where they can see for themselves the destruction of once-beautiful forested hills and productive farmland.

Knowing these facts, a number of landowners in Vinton County were resisting the tempting offers being made for their land. They were attempting to protect the land they had known and loved for many years from being "skinned alive" by the strip-miners. They did not oppose deep mining to the same extent, but they realized that although the coal companies might say they were buying the land or the mineral rights for deep mining, and although they might indeed mine the most valuable deep veins in this manner, there would then be nothing to prevent them from stripping out the veins of poorer-quality coal.

Considering all these facts, the appraiser put a price on the Strasbow farm of about 100 times the price that had been offered for the right-of-way—which would bisect this same land, making it virtually unusable for residence or for mining. Based on this appraisal, there seemed to be ample justification for demanding a very much higher price for the right-of-way.

However, Clovis Strasbow was not satisfied with this. His main concern was to prevent the erection of the line, which his researches had convinced him would be a hazard and would degrade the quality of life along its entire length. Rather than accept even the very high appraisal price for his property, he preferred to force the power company to take account of the dangers it would be creating and redesign the transmission line to be more acceptable.

The day after receiving the court summons, Clovis telephoned many of his neighbors and also the leaders of the protest group in Laurel. One of the landowners recommended a lawyer in Columbus whom he had consulted and who had shown great interest in the environmental aspects of the problem. The following day Clovis and Ada drove the 60 miles to Columbus and retained Philip Morton and his son John to defend them in their case against the Ohio Power Company.

The Mortons were very concerned about the biological hazards created by such a transmission line. They were shocked by the absence of adequate laws in Ohio to protect the citizen against dangers imposed by public utilities which were presumably operated for the public good. They did not accept the contention that it is impossible

M. Post Wolcott for Farm Security Administration

They were trying to protect the land they had known and loved for many years from being "skinned alive" by the strip-miners.

Bill Choyke from *Not Man Apart*, May 1972

to fight a condemnation suit. Precedent or no precedent, they believed that where the safety of the public was at issue there must be a legal way to force the power company to disclose the details of their plan and to obtain a public forum for the facts thus uncovered.

After a careful examination of the possibilities, they decided to invoke Rule 33 of the Ohio Rules of Civil Procedure to require the power company to answer questions concerning the hazardous features of the line. They drew up a list of thirty-two questions asking for technical information on the design specifications of the line; the results of any measurements that had been made by the power company on corona losses, electrochemical products, interference with radio and television reception, and electric shock; and information on any studies that had been made on the long-term biological impact of such lines. These "interrogatories" were filed in the Court of Common Pleas in Vinton County on February 24, and the Ohio Power Company was granted twenty-eight days to answer them.

The power company, however, was totally unprepared for such a response to its suit. Over the years it had developed standard practices for dealing with all the usual forms of legal objection. The demand for technical information was new in its experience. The allowed twenty-eight days passed and no answers to the interrogatories were produced. They would be forthcoming any day, the company promised. In the meantime an overture was made behind closed doors in the law offices for a settlement out of court. "What price would the Strasbows accept for the right-of-way?" But the Mortons insisted that their client was more interested in learning the facts than in acquiring a few extra dollars. This was hard for the power company to believe. It took several hours of discussion to convince their representatives of the sincerity of Clovis Strasbow's conviction.

Still the answers were not forthcoming. Months dragged by, and although there was sound legal basis for asking the case to be dismissed because of the failure of the Ohio Power Company to answer within twenty-eight days, the Mortons advised allowing it additional time. Finally, at the end of May, the Mortons filed a petition asking that the case be dismissed if the interrogatories were not answered in another two weeks. Two days before this deadline the power company produced a document purporting to answer the questions.

This document is a fascinating example of the art of answering demands for factual information with an absolute minimum of facts. Four of the answers were not responsive to the questions asked. Four other questions were not answered at all, objections being raised on technicalities. Five questions which requested quantitative information were answered by qualitative terms, expressing an opinion and revealing no information: "not significant," "relatively little," "similar to," "insignificant." Six questions were answered with flat assertions of "no effect" or "no hazard" without any supporting evidence. The demand for copies of any scientific studies on which the answers were based was denied on a technicality.

In all, only two answers contained any quantitative information. These dealt with specifications for the transmission wire, towers, conductors, and conductor configuration. They revealed, however, the interesting fact that a significant change had been made in the design of the line as it had originally been planned and as it had been used in the first four 765-kv lines. The conductor diameter had been increased from 1.165 inches to 1.38 inches. These larger conductors will produce about half as much corona discharge as the smaller line, resulting in a corresponding reduction in damage to the environment. This heavier construction represents a considerable expense to the company. In late 1969 the engineers of the American Electric Power Company had declared that such "an increase in conductor diameter could not be economically justified."

The change had apparently been made between the time that the formal questions had been asked and the time the answers were finally received. The reason for the long delay was now clear. When faced with the necessity of producing factual information, the power company had rushed out to conduct tests which should have been made years earlier, before any such line had been built. The results of these tests must have convinced the company executives that the line design using the small conductors was indeed too high in corona loss, that it produced levels of audible noise and electrochemical pollution which would not pass the critical inspection of informed and concerned citizens.

This change to a more conservative design of transmission line would never have occurred without the efforts of the little protest group at Laurel that originally brought the issue to public attention, or without the efforts of Clovis Strasbow, who had organized his own field tests and had been willing to stand up against a giant company rather than accept a large settlement for the right-of-way. And finally, it would not have occurred without the legal help of the Mortons, who were able to force the power company officials to address themselves to these environmental concerns. Because of the action of these few people, two to three thousand miles of 765-kv line projected for construction in the next few years will be built with conductors that should produce only half as much electrochemical pollution. This certainly represents an important victory.

The battle, however, is not won. Even with these improvements, this line will be a very unpleasant and polluting installation. The electric field effects and shock hazard will not be improved, and visually the design remains just as horrendous. Furthermore, under the terms of the standard right-of-way agreement, the company can make any changes it likes later in the construction and operation of the line. They can run higher voltages over these larger conductors, thereby completely undoing the beneficial effect of the more conservative design. They can build taller structures and add more transmission lines over this same strip of land. Once the right-of-way has been appropriated, the landowner has no protection against these changes.

Clovis Strasbow's appropriation case is still pending. If he takes it to court, the expenses involved might exceed the amount of damages paid for the right-of-way. Clovis would be opposing a battery of top lawyers and "experts" hired by the electric company. A multibillion-dollar business backed by legal precedent is a formidable coalition for an individual to fight.

12

Earthspace Is Precious

Ever since the memorable day in 1967 when two American men looked back from their spacecraft and saw our planet in perspective for the first time in human history, a new awareness has been taking shape in the American consciousness. There is a dawning appreciation of the exceptional endowments of this planet, which make it a congenial environment for the tender web of living things. There is a new understanding of the finely adjusted balance of all the processes taking place in the swirl of ocean and cloud that wraps the earth like a luminous blue veil. Most important of all, there is the sudden realization that our home in space is *small*. Until we went away from the earth and saw it set against the immense spaces of the universe, this home of ours had seemed limitlessly big, just as to the child the house where he grows up seems very large. After he has been out in the world for a while and returns home, he is surprised to find how little it is.

As we put into orbit satellites that beam messages around the earth in seconds and see on our television screens views of the earth that encompass three continents in a single shot, it is impressed upon us that our home is limited both in space and in time. Earthspace is precious. There will never be any more of it. It can only be stretched by learning to use it more wisely.

Each human being expends a certain amount of body energy just to stay alive and requires a certain amount of earthspace for raising the food that provides this body energy. To these minimum amounts of energy and space there must be added each person's share of the additional energy expended by the technology that supports him, the space occupied by the factories, stores, streets, and transportation systems that distribute the products of the technology. Per capita consumption provides a rough measure of each person's share of this supporting technology. Thus the impact of a population on earthspace is affected by both population size and per capita consumption. In some parts of the world today the population size is the variable that is changing most rapidly, but in the United States the rate of change of per capita consumption far outweighs the increase in numbers.

The population of India is more than twice as large as the population of the United States and is growing at a faster rate. But on the other hand our gross national product (in constant dollars) is doubling every twenty years and our use of electricity is doubling every ten years. One American uses 95 times as much electrical energy as one Indian. In terms of electrical consumption our population is equal to that of 20 billion Indians today; and, if present growth rates continue, by the year 2000 it will be equal to that of well over 100 billion Indians.

Unfortunately, a belief in the expanding economy is deeply entrenched in the minds of American businessmen. Growth in both population and per capita consumption are thought to be essential for continuing profits. Although the statistics of the demographers demonstrate beyond a shadow of a doubt that sustained population growth at the exponential rates occurring in the 1950's would soon result in a world where there is standing room only, businessmen have not really accepted the concept of a stabilized population. They are even more reluctant to accept the need for a reduced rate of economic growth.

Yet it is obvious without working through all the frightening statistics again that a population that doubles every thirty years is soon going to snuff out all competitive living things and eventually smother of its own weight. It is equally obvious that an industry that doubles its production *every ten years* will rapidly overcome any competitive methods and the products it creates will crowd out everything else in the earth system.

The doubling process is a truly awesome phenomenon, but it begins so innocuously that its consequences are not immediately apparent to people who are unaccustomed to working with large numbers.

To illustrate how easy it is to fall into this trap, George Gamow related an old legend about King Shirham of India. This King Shirham had a grand vizier, Sissa Ben Dahir, who had invented and presented to the king the game of chess. King Shirham, wishing to reward his grand vizier, asked him to name his wish. The clever vizier knelt in front of the king holding the chessboard and said, "Majesty, give me a grain of wheat to put on the first square of this chessboard, and two grains to put on the second square, and four grains to put on the third, and eight grains to put on the fourth. And so, oh King, doubling the number for each succeeding square, give me enough grain for all sixty-four squares of the board." Smiling at this seemingly modest request, the king called for a bag of grain and servants to count it out, but it soon became obvious that many more bags of wheat were needed. In fact, all the wheat in India could not fulfill the promise. To satisfy Sissa Ben Dahir's request would have required the whole world's wheat production for some two thousand years.

We can think of earthspace as the squares of the chessboard and imagine what will happen with each succeeding decade when the power plants, transmission lines, and switching yards are doubled. The rapidity with which all available land would be used up is vividly demonstrated by a calculation involving just power plants alone. "Suppose," said the Committee for Environmental Information, "that all electric power is to be produced by modern 1,000-

megawatt power plants and that each requires an area of 1,000 feet on a side. If all of the country's power needs were presently being met by 300 such large power plants, in less than twenty doublings—that is, in less than two centuries—all of the available land space in the United States would be taken up by such plants. Not the available fuel resources or water for cooling—just physical space. This does not leave room for transformers and transmission lines, let alone people."

Rights-of-way for power lines are far more demanding of land resources than the generating facilities themselves. For every plant occupying a few hundred acres, thousands of acres of right-of-way are required to distribute the electricity. Thus our dwindling land resources and the many competing new needs for earthspace bring into focus a number of important questions. Should electric companies be allowed to site their plants hundreds of miles from the principal areas they will serve and appropriate rights-of-way through farmland and forest to transmit the power? Should they be allowed to flood thousands of acres of land to facilitate the generation of electricity? And if so, which acres?

Intelligent planning of land use is needed to achieve maximum utilization of our most important natural resource. Years of study and artistic talent are expended on the organization of our interior spaces; yet we continue to allow completely haphazard and destructive utilization of our exterior spaces. In the meantime, long-range planning for optimum land use becomes daily more difficult and expensive. Land destroyed by strip-mining cannot easily be returned to productive farming, nor can land covered by concrete highways. Land crisscrossed by power lines cannot be used for airports. It is spoiled for recreational areas or building sites.

Utility Corridors and Multiple Use of Rights-of-Way

One possibility for improving the efficiency of land use is the establishment of common rights-of-way for utilities; gas and water pipes, telephone and electric lines. By consolidating these supply and communication lines into common corridors, the total use of land for rights-of-way could be reduced several-fold. However, the design of electric lines presently being constructed makes utility corridors impractical. Extra-high-voltage overhead lines create such a hazard of electric shock that the installation and maintenance of other utility lines becomes a major problem.

A recent article in a professional magazine for pipeline engineers contains elaborate instructions for protecting personnel against the possibility of lethal shock when installing a pipeline under an energized high-power line. "When circumstances are right," the author warns, "the magnitude of the induced potentials can be sufficiently great as to represent a hazard to a person making contact with the pipe steel at an above ground appurtenance. This hazard can exist during the occurrence of a transient a-c condition on the electric transmission system, i.e., a lightning stroke causing arc-over at an insulator or a switching surge occurring when the line is brought into or removed from service. The hazard also may exist during steady state a-c system operation, with the greatest effect being felt during the time that the pipeline is being constructed and is above the trench."

These and related problems such as corrosion of the pipe from induced potentials and electromagnetic interference with communications systems, make it unlikely that other utilities would voluntarily elect to share rights-of-way with high-voltage overhead transmission lines. The concept of joint rights-of-way or utility corridors makes very good sense from the point of view of land use. However it is not feasible as long as the electric companies build their transmission lines in such a way that dangerous currents can be induced.

Another recommendation, made by the Department of the Interior and the Department of Agriculture, is the use of transmission line rights-of-way for public recreation: golf courses, picnic parks, wildlife sanctuaries, equestrian or bicycle paths. But the possibility of electric shock from conducting objects under the extra-high-voltage lines would make the use of these areas unpleasant, if not positively dangerous. Birds and animals are frightened by the electric field. Ornithologists have observed that birds will not fly near the new extra-high-voltage lines. Animals give them a wide berth whenever possible. Horses wearing bridles are annoyed and chafed by the currents induced on the metal bits in their mouths. Thus the areas under these lines will hardly be useful as wildlife sanctuaries or bridle paths.

Furthermore, the hazards of electric shock will increase as the utility companies build lines carrying 1500 kv or even 2000 kv. H. C. Barnes of American Electric Power Company, speaking at a symposium of international power engineers in Sweden in August 1971, suggested that the problem of induced shock from ultra-high-voltage lines might be solved by purchasing and fencing the rights-of-way. "This solution," he said, "may be more economical for the power companies than increasing ground clearances."

Fencing rights-of-way for transmission lines would represent a major usurpation of land. Conservative estimates project the construction of 300,000 miles of new transmission lines in the United States by the end of this century. Fenced-off corridors for these lines would mean at least five million acres requisitioned for the transport of electricity. In addition, the fencing would divide farm properties, isolating sections of land from barns and machinery. This division of property would effectively take much more land out of production than the right-of-way itself.

Under our present laws electric companies are permitted to erect lines that make the land beneath them virtually unusable. The right of eminent domain gives them the power to condemn and fence off millions of acres just because this solution may be *more economical than increasing the height of their lines.* With the increased pressure on land resources and the increasing use of land for non-productive purposes, a generation from now such rights-of-way, surrounded by high fences, sprayed with brush killers, and posted with "Enter at Your Own Risk" signs, may be the principal hiking trails and recreation areas available in America.

Land-Use Issues

There are a number of other ways in which the electric power industry is making new demands on land resources because the use of land provides a cheap solution to a technical problem. Artificial cooling lakes, which are one answer to the threat of thermal pollution from power production, are less expensive to build than cooling towers. In some locations they have aesthetic advantages; however they do require large areas of land. A surface of one to two thousand acres is needed for each 1000 megawatts of generating capacity. In many cases this land is very valuable for other purposes.

Commonwealth Edison's new nuclear plant near Seneca, Illinois, provides an example of the land-use issues that result from the appropriation of land for cooling lakes. The site chosen for its lake covers 4480 acres of fertile prairie land whose flat contour makes it very economical to convert into a shallow lake with low retaining dikes. Grading and landfill expenses will be minimal. On the other hand, the flatness and fertility of the black prairie soil also make the land ideal for growing corn by the continuous cultivation method currently used throughout this prime farm region.

Property owners are vigorously opposing appropriation of their land for the cooling lake. They point out that just to the west of the proposed site lie thousands of acres of land that was strip-mined and left valueless for agricultural purposes. This land is badly cut up, with steep banks and gullies. The fertility of the soil has been destroyed by burial of the topsoil and by erosion. But Commonwealth Edison protests that it cannot use this land for the cooling lake because the additional grading and leveling would add considerably to the expense.

The legal question to be decided in this case is simply the value of the property appropriated; but the real issues go much deeper. Should a power company be allowed to appropriate land that is valuable for farming because it is cheaper to use this land than to use land that has no value for other purposes? Should a company be allowed to build a cooling lake at all? There are cooling towers that do the same job; and some of the most recent tower designs are not very objectionable from an aesthetic point of view. There is a new type of tower only 60 feet high which can easily be concealed by planting. Towers of this kind are about 600 feet in diameter. They encompass only half an acre instead of four and a half thousand acres.

Similar issues arise in connection with pumped-storage reservoirs such as the one planned for Storm King Mountain on the Hudson River. In this case the installation threatens to destroy recreational and aesthetic values. The decision in these land-use conflicts should not be based on price alone; the advantages and costs to society should be impartially evaluated. In some instances the use of land *may* provide the optimum solution. In other cases pressure should be brought to bear on the utility company to use the slightly more expensive solution and develop alternate technologies that would provide better options for the future.

Without enlightened long-range planning, our earth will inevitably grow more congested as more people expend energy at a higher per capita rate and are compacted into less effective earthspaces. According to estimates of the electric industry itself, in the year 2000 we will be using twenty-five times as much power as we were using in 1950. The environmental impact of this power production and transmission would be the same as if our per capita consumption of electricity had remained constant and our population had grown from the 152 million of 1950 to four billion in 2000. Unless a progressive ongoing technology finds ways to reduce dramatically the environmental impact of producing this electricity, air and water pollution will escalate. Astronauts of the next century may be able to see the difference as they look back at our planet. The luminous blue and white veil of ocean and cloud will have turned into a heavy shroud, gray and dirty at the edges.

13

A Question of Power

During the spring and summer of 1971, several newspaper articles about the 765-kv lines appeared in local Ohio papers, and a television news program devoted a quarter of an hour to the subject. This publicity attracted the notice of the attorney general of Ohio, who initiated some inquiries into the environmental effects of the extra-high-voltage transmission lines. A letter was sent to the American Electric Power Company asking for scientific evidence of the safety of these lines.

In due time the company replied by sending several of their most important engineers and lawyers out from New York with the results of a "token experiment." The field measurement data they showed the attorney general consisted of seventeen readings of total oxidant concentrations near extra-high-voltage lines. Within the range of accuracy of the instruments, only four of these readings were higher than oxidant concentrations away from the line. However, only five of the seventeen had been made under 765-kv lines in weather conditions when corona discharge could be expected to occur, and *none* of the measurements had been made when the lines were fully energized to the voltages they are designed to carry (765-800 kv).

This experiment did not convince the attorney general. A short time later he sent a member of his staff to call on some of the people who live under the line operating between Piketon, Ohio, and

Louisa, Kentucky. Clovis Strasbow acted as guide. The trip made a deep impression on the man from the attorney general's office. He was surprised to discover the various discomforts imposed on the people living in the path of this line. Most of all, he was shocked by the discriminatory manner in which the electric company responded to the complaints of the landowners. Those people who had money and local influence had received fairly good attention. But the complaints of several very poor families had gone unanswered. Although the company was theoretically responsible for repairing any destruction of property caused by the installation of the line, many evidences of damage were still obvious three years after the work had been done. The heavy bulldozers that cleared the right-of-way had cut steep banks, destroyed drainage systems, and left large areas bare and eroded. One man had to shovel clay and mud out of his front yard every time it rained. All the residents near this line lived with a constant sense of its threatening presence, the fear of electric shock by some inadvertent action, and the irritating noise that invaded their homes and destroyed their enjoyment of their yards and gardens.

In view of his staff member's report, the attorney general requested the Ohio Power Company to delay energizing the 765-kv line it had just completed between Burger and Lima. For several months the company complied. But eventually the pressures for energizing this line became so great that the attorney general was no longer able to effect a postponement. Although he questioned the environmental consequences of such a line, he did not have the power to compel a more adequate environmental study. In March the following news item was released by his office:

> Some months ago, we expressed concern that there may be adverse environmental consequences when a 765KV line is fully energized. A member of my staff, Barry Smith, has been looking into the matter for me. We now know audible noise can result, ozone gases may be given off, and radio and television reception can be interfered with, among other possible effects. Thus far, we have no assurance that steps will be taken to eliminate such possibilities in connection with the EHV line which is being energized today, and we have serious questions as to

> whether the future construction and full energizing of EHV lines can be accomplished within the guidelines of the national environmental policy as stated in the National Environmental Policy Act of 1969.
>
> No State or Federal agency has authority over the location and operation of the EHV lines except the Army Corps of Engineers, and then only where the lines cross a navigable river. [There are currently permit requests pending for additional EHV lines which would cross the Ohio River at the southern and southeastern points of the state.] We are, therefore, asking the Corps to undertake a study of possible environmental consequences in regard to the EHV lines.

At the present writing, the Corps of Army Engineers has not undertaken these studies.

Another government agency, the U.S. Forest Service, had also become concerned about the extra-high-voltage lines, fearing that forest growth in the eastern United States might be injured by continuous exposure to the electric fields and the electrochemical products of corona discharge. In order to evaluate these hazards, the Forest Service planned to undertake a careful in-depth investigation of long-term effects on various types of vegetation. The research they proposed would have taken three years and cost about $200,000. Unfortunately, no government funding for this study was available. The Forest Service then attempted to interest the Ohio Power Company in financing the study, but the company was not interested in such a careful scientific evaluation of the problem. Instead it hired a consulting agency to produce the quick answer it wanted. This field study was completed in three days instead of three years; it consisted of making measurements of oxidants for a total of nine hours under a line designed to carry 765 kv, but no record was made of the voltage actually being run at that time.

It is obvious that an environmental investigation funded by a company that has a vested interest in the results of the investigation is apt to be self-serving. Federal funding appears to be the only way of obtaining an impartial study, but federal funding is not always available even to a government agency that is convinced of the need for the research. "The questions raised by these lines," said the attorney general, "are sufficient to merit the utilization of the federal government's research facilities to determine the full environmental impact of EHV transmission lines."

In the meantime two lines continue to operate in Ohio. Rights-of-way are being bought for two more. If the attorney general of Ohio is powerless to prevent the electric companies from building and operating questionable equipment, what chance does the ordinary citizen have? What chance does Clovis Strasbow have? Or Kiziah Hough? Or Taterbug Brown?

14

What Are the Alternatives?

The hazards and environmental damage associated with extra-high-voltage lines might be justified if they were necessary, but they are not. Sufficient energy for all our needs can be carried in less destructive ways. However the technological improvements necessary to do the job right can only be achieved by a greatly stepped-up research and development program.

For at least a decade now the electric industry in the United States has not been investing sufficient time or money in research. Fifty-one of the two hundred and twelve major electric companies, according to the Federal Power Commission, did not spend a cent on research and development in 1968. In the industry as a whole, less than one cent of every dollar of gross revenue was spent on research in 1971. The consequence is that we have been steadily falling behind the other advanced countries of the world in pioneering designs for electrical equipment. Creative new solutions for power transmission and generation problems are actually at hand today, but almost all of these innovations have been invented and developed in other countries. The relatively few contributions made in the United States were made not by the electric power industry, but by the aerospace, chemical, electronic, oil, and gas industries. As far as our power industry is concerned, design innovations are judged on how well they immediately serve the prime objective of delivering maximum power at minimum cost. If the objective was, instead, to produce and transmit power with maximum economy of natural resources and minimum impact on the environment, then a number of alternative solutions would be attractive.

An overall look at our power transmission needs suggests that the ideal solution would involve the establishment, over a period of time, of a planned countrywide network of high-voltage transmission lines which would provide maximum efficiency of power transmission with losses so minimized that electricity could be routed by a longer circuit without serious loss of power. In this manner peak loads could be balanced and shortages could be made up from other, relatively remote, parts of the system. Such rerouting of power is not practical over long distances in the present system. The older, lower voltage lines lose a considerable amount of power per mile through heating losses. The new extra-high-voltage lines have lower heating losses but are so designed that the corona losses to the atmosphere are considerable; these losses make it economically necessary to send the power in the shortest straight-line route from plant to consumer. Although a grid of minimum-loss lines would be more expensive to build than the present marginally designed lines, over the years the investment in a system of this kind would pay back dividends in reliability and economy of operation. By averaging out peak demands it could result in important savings in generating facilities; and at the same time it would serve the important function of conserving the quality of our environment.

A larger investment in equipment, more expensive towers and conductors, would make it possible to build overhead high-voltage lines that would carry electricity with lower losses per mile than any lines in service at the present time. Such low-loss lines would produce less corona discharge and cause less radio and television interference. The visual pollution could also be reduced by more graceful tower designs —already available today—and by planning the routes to avoid cutting straight swaths through scenic areas. These improvements would not require further research expenditure. The principles are already known.

Underground Transmission

However, the most effective way of reducing the environmental impact of power transmission would be to bury all or most of the lines. Underground transmission would solve many problems at once. Underground lines would not mar the landscape or electrify the air; they would not attract lightning, and would be much less vulnerable to storm damage. They would never cause the build-up of static electric charge. Their installation would require the destruction of fewer trees, and once the earth was replaced vegetation could be allowed to grow back across most of the cleared strip.

Electric lines have been buried throughout urban areas and many suburban areas for years. The power is carried in cables laid in trenches three and a half to five feet deep. The most commonly used underground cable has copper conductors, each conductor consisting of a bundle of several hundred strands of copper wire elaborately interwoven. This design requires meticulous care in manufacture, and splicing the lengths of cable together is a very time-consuming operation. The entire conductor is wrapped with insulation paper kept saturated with oil under pressure to improve its electrical insulating properties.

These cables are made as small in diameter as possible so they can be handled more easily. The distance from the outer surface of the conductor to the sheath is usually on the order of one inch. With an alternating current, there is a flow of so-called "charging" current between the conductor and the sheath and this current represents an electric loss. Its magnitude decreases with the thickness and effectiveness of the insulation; it increases with the length of the cable and with the voltage. The energy in these charging currents, and also the losses caused by resistance to the current flow, are dissipated in heat that must be absorbed by the surrounding earth. Removal of this heat creates problems that are particularly troublesome in ac underground transmission. These facts put a natural limit on the length of underground line that is economically practical for the transmission of ac currents at high voltages. At low voltages like those used on most distribution lines (4000-35,000 volts) the losses from charging currents are not important.

The demanding tolerances required in manufacturing and joining together lengths of this copper cable add to the expense of the line. Pulling the cable into pipes and providing manholes for splicing and repairing sections of cable also represent a major part of the cost of underground installation. These problems vary enormously with the type of terrain and other obstacles such as buried sewers and gas lines that exist in the area. Threading power cables through an existing maze of other pipes and lines is a costly undertaking. In spite of these difficulties, however, new work methods for placing distribution lines underground have resulted in major economies in the past decade. Within new residential subdivisions underground installation costs only 1.5 to 2 times that of overhead lines.

Because there has been little necessity in this country for putting major transmission lines underground, the art of transmitting extra high voltages by underground cable is in a much more rudimentary state of development than the lower voltage methods suitable for distribution lines. The techniques currently in use are essentially adaptations of the lower voltage technology and are costly and inefficient for high voltages. An underground cable of the oil-insulated type carrying ac current at 345 kv has so much charging current that almost all the current-carrying capability of the line is wasted in a distance of about twenty-six miles. Lines carrying 345 kv are not installed underground for spans of more than fifteen miles.

These limitations can be overcome by transmitting direct current. With dc there is no charging current and, therefore, dc underground lines can be run for long distances. Dc cables are less expensive than comparable ac cables. They require one less conductor per circuit and they carry a larger current for a given size. The insulation can be thinner because the heating problem is much less important with dc cables. However, high voltages are most efficiently obtained with ac current and most electrical appliances are designed for ac. Therefore, when high-voltage dc transmission is used, ac current is usually converted to dc, transmitted, and reconverted to ac before it reaches the consumer. To make these conversions, rectifiers and inverters are necessary, and these are expensive. However, the development of solid-state technology is resulting in significant price reductions for rectifying and inverting equipment. Further research and mass production of the units will undoubtedly continue to improve these economies.

In order for dc transmission of any kind to be competitive in price with ac at the present time, large blocks of power must be moved long distances. Dc is especially advantageous where sections of line are to be put underground. Both of these conditions apply in many locations in our country; but our utility companies are clinging to the old solutions. They are attempting to avoid any underground transmission. Where such a run is absolutely necessary they use the existing ac technology, even though the cost for this underground stretch is very great, rather than put the money and effort into developing dc technology. No high-voltage underground dc cables have been installed in the United States, but such cables are in use in other countries for long underwater crossings. In New Zealand a 500-kv dc line with a total length of 385 miles has been installed between North Island and South Island. Twenty-five miles of this passes underwater across Cook Strait. Another dc cable links the electric power systems of France and England under the English Channel. Another cable 60 miles long connects Sweden and Gotland.

Several new types of underground cable have been designed, utilizing different conducting and insulating materials. Cables have been filled with polyethylene, polypropylene, and various other synthetic polymers, which have insulating properties far superior to oil-impregnated paper. Cable with conductors made of sodium has been used in several installations in the United States. This cable is relatively lightweight and flexible, but, although sodium cable is considerably less expensive than the commonly used type of underground cable, there is not enough demand to justify commercial production. Union Carbide announced in 1971 that it was discontinuing its manufacture.

Union Carbide has also been active in the development of a transmission technique that offers maximum efficiency for power transmission—superconducting cables. When the temperature of a metallic conductor is reduced to near absolute zero (—459°F), the metal becomes an almost perfect conductor. Heating losses are so dramatically reduced that the conductors can carry a very heavy current load. The low temperature is maintained by a refrigerant such as liquid helium pumped through the conducting pipe.

Superconducting systems are most easily adapted to dc transmission. In fact, a superconducting dc line would offer the ultimate in transmission without loss; and cables to do this job have been designed. One detailed study, made some time ago at the Thomas J. Watson Research Center of International Business Machines Corporation, estimated that a 600-mile long superconducting dc line with a capacity of 100,000 megawatts would cost considerably less than comparable overhead transmission. However the potential advantages of this system have been neglected in this country. Great Britain has been the leader in the development of supercooled dc electrical equipment. Russia also is pioneering in this technology. A supercooled dc line will deliver power to an aluminum plant from the Bratsk Dam in Siberia.

In the early 1960's, scientists at Union Carbide began a search for a superconductor that would handle ac. By 1967 they had discovered a conductor made of very pure niobium that was capable of carrying ac with extremely low losses. Theoretically, one 345-kv superconducting line twenty inches in diameter would carry enough power for all of New York City. Twenty-two conventional cables ten inches in diameter would be required to carry the same amount of power. Such superconducting cables installed underground would offer price advantages over conventional underground cables. The idea looked so promising that the Edison Electric Institute financed a study that culminated in 1969 with the demonstration of a twenty-foot-long superconducting cable. However, when Union Carbide proposed to follow this study with an $8-million pilot program to build a short superconducting line for field tests, the U.S. electric companies failed to put up the money for it. In 1971 they undertook a less ambitious project financed jointly by Edison Electric and the Department of the Interior. This $2.11-million, three-and-a-half-year study is devoted to constructing and testing a laboratory model. Although several more years of engineering work will be required before superconducting lines are commercially feasible, scientists at Union Carbide see no fundamental obstacles to the successful use of these lines.

A modification of the supercooled line has been developed by General Electric, using aluminum underground cable cooled to a minus-320-degree level. This cable can handle ac voltages up to 435 kv and large current loads, so that the power transmitted could be seven times greater than the highest-power underground cables now serving metropolitan areas. Cables of this design represent a compromise between the present oil-insulated systems and the superconducting cables made of niobium and cooled to absolute zero. They are less expensive to build than the niobium cables, but they operate with higher electric losses. A forty-foot section of this "cryogenic" cable has been built and tested; the next phase is to design a commercial underground system.

A different concept in transmission technology has been used in Europe and is feasible today for ac current at voltages as high as 345 kv. This is a gas-insulated cable, using the inert non-toxic gas sulphur hexafluoride (SF_6), which has such superior insulating qualities and such great stability that it fully insulates a conductor from the outer tube. Aluminum tubes carrying the conductors and the insulating gas can be placed side by side on low pylons, in open trenches, or underground. Such installations represent an enormous economy of space as compared with the minimum separation of 10 feet required by bare metal cables. These lines can be made completely enclosed, grounded, and safe. Lines carrying enough power for an entire metropolitan area could be laid under highway median strips where they would cause no visual offense, no shock hazard, and no air pollution. SF_6-insulated lines and substations are at present about 20 to 100 per cent more expensive than overhead installations (not counting land-cost savings), but aggressive development of this technology could rapidly close the cost gap.

Again, the United States utility companies have followed rather than led in the development of the new technology. Companies abroad have already dotted the European landscape with SF_6 substations which use up only one-twentieth of the land required for conventional substations. Although several small test installations are presently being planned or built by U.S. utility companies, the installations are designed on principles developed abroad, and one of them has been contracted through a European firm.

The SF_6 technology does offer a practical and viable alternative to extra-high-voltage overhead transmission. Steadily rising land costs will reduce the cost differential between the two systems. If our utility companies were required to underground a significant portion of their new transmission lines the SF_6 would soon become economically competitive.

Alternative Technologies

The transmission problem could be solved indirectly by the development of power sources to generate electricity at the point of consumption. Fuel cells which convert gaseous fuel directly into electricity perform this function and have a number of other advantages as well. Fuel cells convert energy into electricity much more efficiently and with fewer undesirable by-products than the steam-turbine conversion process. Therefore, their use would conserve natural resources and reduce the thermal and atmospheric pollution caused by conventional generating plants. Even small units suitable for houses or apartment buildings are efficient enough to warrant their installation at the point of consumption, thereby avoiding electric transmission problems altogether.

Fuel-cell technology, which was pioneered during the 1930's, was vigorously pursued during the early 1960's in order to perfect a lightweight, efficient power source for space vehicles. These cells, developed for the special needs of the space program, are too expensive to be commercially feasible for home use. However, the theory of fuel-cell design is well understood and a relatively small investment in engineering work could result in fuel cells that would be economically practical for small installations as well as for large-scale central generating stations. Unfortunately, government support for fuel-cell development has dropped to a fraction of its former figure, and although some research is being conducted by industry, particularly the gas companies, the total effort is very inadequate considering the important environmental advantages of the fuel-cell system.

Fuel for these cells can be natural gas or gas produced from coal, oil, or organic wastes. Coal is the most abundantly available energy resource in this country, but burning it as fuel poses many environmental problems. It creates soot, fly-ash, sulphur fumes, and nitric oxides. In the present state of the technology these elements cannot be removed completely from the stack gases, and the existing techniques for partial removal are expensive. As a consequence, much of our coal, especially the high-sulphur type, is unsuitable as fuel.

However, industrial processes have been in existence for the past thirty or forty years for the conversion of coal to manufactured gas. During the gasification process the sulphur content can be removed and a relatively clean fuel produced. The burning of gas does not create soot or fly-ash. Because the technology for making and using coal-gas has been with us for years, a minimum of time, effort, and expense would be required to convert our most abundant source of energy into our cleanest source of energy.

Gasifiers are already used extensively in Europe, where fifty-eight installations are now operating. Only two pilot plants have been built in the United States. One of these experimental plants, built to test the "Hygas" process, is part of an eight-year program financed by the American Gas Association and the Department of the Interior. The plant is expected to be able to produce 1.5 million cubic feet of gas a day and to make gas equivalent in quality to natural gas.

A form of gas can also be made from agricultural and urban organic wastes by a process of fermentation. In the absence of oxygen, microorganisms transform the organic matter into a gas which can be easily cleaned of pollutants. The solid-waste produced annually in the United States could yield half again as much gas as our current natural gas consumption. Although the conversion equipment would represent a considerable capital expense, it would be performing two important functions simultaneously: disposing of the polluting organic wastes produced by our cities, farms, and food-processing plants and, at the same time, making the chemical energy stored in these organic compounds available as a renewable energy resource.

Gas manufactured by these methods can be used to fuel electric power plants at the sites where it is produced. Cost estimates indicate that power plants using coal-gas, for instance, would actually be less expensive than conventional coal-fired steam boiler plants with good stack clean-up equipment to remove soot, fly-ash, and polluting chemicals. However, the electricity generated in this manner must then be shipped by high-voltage transmission lines across the country to consumers.

Alternatively, the gas itself can be transmitted in underground pipelines to population centers. These underground gas pipes can be installed much more economically than the oil-insulated underground electric cables in current use. It is estimated that the distribution costs of gas are only about 20 per cent of the cost of distributing electricity through overhead cables. During off-peak hours some of the gas can be stored underground in liquid form under pressure. It can also be stored and shipped in pressurized containers. Some of the gas can be used directly for heating purposes and the remainder used to generate electricity in small power plants or in individual fuel cell installations.

This system of transporting the gas to the point of consumption has a number of advantages over the transmission of electricity. The ability to store gas makes it possible to even out annual and diurnal variations in energy demand. Electricity, on the other hand, cannot be effectively stored in any quantity. The higher efficiency of using gas directly for heating results in a significant saving of natural resources and a reduction in the amount of waste heat added to the environment. Finally, the transmission of gas is much more economical than the transmission of electricity, and the gas can be piped underground without destroying the landscape.

It is not surprising, however, that the big electric companies such as Westinghouse and General Electric are pushing the system based on converting coal-gas immediately to electricity, since this system gives the electric industry the biggest slice of the energy market. It sustains a high demand for kilowatts as well as for the various appliances that operate on electricity. The large electric companies are currently seeking government support for this system. The gas industry, on the other hand, favors the alternative that utilizes gas transmission technology and appliances using gas. The federal government's decision on the amount of funding for these two competitive systems may well determine the nature and effectiveness of the technology that will bring energy into American homes in the 1980's.

Another promising energy source for fuel cells is hydrogen, which can also be made from fossil fuels and can be transported efficiently in pipes, using basically the same technology developed for the transmission of gas. The burning of hydrogen produces only water as a by-product. It is an ideally clean fuel. This method of transmitting energy has already been put to commercial use in Germany, where hydrogen made from crude oil is transmitted in underground pipelines to large steel mills. By the summer of 1970, 125 miles of hydrogen lines had been installed in the Ruhr.

Hydrogen can also be produced by passing a current of electricity through water. Using this conversion process, Dr. Derek P. Gregory of the Institute of Gas Technology in Chicago has suggested a very interesting new concept in energy distribution. He proposes that electricity at the power plant be used to decompose water electrolytically into hydrogen and oxygen. Hydrogen would then be transmitted by underground pipeline to the point of consumption. As with natural or manufactured gas, some of the hydrogen could be stored under pressure in liquid form in underground reservoirs, or alternatively in pressurized containers. Some of the hydrogen could be used directly in fuel cells for home heating or as a fuel in industrial processes. Some could be converted to electricity in small local stations and distributed to the consumer.

Dr. Gregory believes that the system he proposes would be no more expensive and might actually offer economic advantages over the present methods of generating and transmitting electricity. He calculates that the capital costs of installing a hydrogen-delivery system, including the electrolyzer plant, would be competitive with the costs of installing an EHV transmission line and its transformer stations. Although the hydrogen might cost about 60 per cent more to transport than gas (because of their physical differences), this still means a very nominal transmission cost per mile. When the fixed expense of conversion to hydrogen is added, it appears that energy delivered to the consumer in the form of hydrogen would be cheaper than electrical energy at distances of more than 500 miles. For longer distances much greater savings would be effected because the transmission itself is so economical.

The economy in transmission would make it possible to locate large atomic power plants far from metropolitan centers, thereby providing an extra safety factor against the eventuality of nuclear accidents. It would make possible the interchange of energy from distant parts of the system to compensate for unusual loading conditions and to even out peak demand. All these advantages would be obtained without visual damage to the intervening landscape.

The technology for implementing a hydrogen system would require considerable developmental work. However the basic concepts are well understood, and satisfactory models for the various components are actually at hand today. Reliable and highly efficient fuel cells and rocket engines burning hydrogen and oxygen were developed for the space program. With a reasonable investment in time and money, the technology that took us to the moon and has put sophisticated instruments into orbit around Mars could help us make our own planet a better place to live.

The crisscrossing of the planet's surface with towers and cables is defacing our broad vistas and creating a visual tick-tack-toe across little scenes of hill and field and rock outlined against the sky. None of us would build a house or even a mobile home with all the plumbing lines and electric conduits strung out in full view and cutting diagonally across the living room. Of course it would be cheaper to do so, but we would rather pay a little more to create clean, uncluttered living spaces in our homes. We Americans are the richest people in the world; we can afford to preserve uncluttered beauty in our outdoor living spaces as well.

The telephone company is embarking now on a program to bury almost all of its lines. The company estimates that in the year 2000 over 90 per cent of the lines will be underground. Visually, however, telephone lines are much less offensive than the enormous electric lines. So while the telephone company is investing money in improving the attractiveness of our countryside, the electric utilities are moving full speed ahead with plans to invest millions in new overhead lines that will desecrate the landscape.

In February 1971, the New York Public Service Commission proposed rules to require both electric and telephone companies to put their existing overhead local-service lines and equipment underground throughout the state. The proposal called for the companies to earmark 2 per cent of their gross revenues each year for the following year's conversions. Some of this cost would be borne by the consumer in increased rates. The commission also studied the possibility of requiring that transmission lines be placed underground.

Estimates of the expense involved in burying all new lines—both distribution lines and transmission lines using the old oil-insulated design—vary enormously. In some cases the expense is estimated to be twice that of overhead lines and in other cases 15 to 20 times higher. Perhaps the best way to weigh the advantages of an underground transmission system versus the cost is to consider how much it would cost each householder in terms of increased electric bills.

In 1966 the Federal Power Commission estimated that if 10 per cent of all transmission is underground by 1980, the average cost of power to the retail consumer would be increased 11.8 per cent. If 20 per cent of all transmission is underground by 1980 the cost increase would be 21.5 per cent. However, at the same time the FPC estimated that improved technology in the power industry in general would cut the cost to the consumer by 27 per cent. Therefore, according to these figures, a quarter of our transmission lines could be buried over a fourteen- or fifteen-year period and the cost would be totally absorbed by the technological economies. At this rate all lines could be buried within 60 years for no cost increase to the homeowner.

Furthermore, if the utility companies were required to bury many of their lines, the technology of underground installation would be aggressively developed. The newer and less costly ways of doing the job would be exploited so that this conversion time would be reduced. The sooner we start on this program, the sooner we will reach the day when Americans can enjoy again an expansive view of land to an unbroken horizon.

Clutter is disturbing to the spirit. It destroys the serenity of broad vistas; in crowded areas it is intolerable. As the years go by our living space on earth will inevitably become more restricted, but the feeling of spaciousness can be augmented even in restricted areas by an orderly and harmonious arrangement. Hiding the ugly mechanics of our civilization will make our living space seem less crowded. It will give us back views of tranquil countryside and space for the expanding spirit.

I once heard a twelve-year-old girl describe her pleasure when during a trip to our plains states she had seen for the first time a clean sweep of prairie land to an unbroken horizon. "Just earth and sky," she said, stretching out her arms in a joyous gesture of freedom. "There were no buildings or people or things to mix you up, so you could really feel how the earth is and see the curved shape of it against the sky."

Alternative Power Sources

The solution of the two key problems of low-loss transmission and effective storage of energy would immediately make the utilization of certain alternative power sources practical. There are a number of energy sources that are free and constantly renewed by nature. Some are abundant, and most could be utilized without the addition of any polluting elements to the environment. However the inconvenient location and extreme variability of these natural resources has made them difficult to develop without good transmission and storage systems. Within the very rigid framework of our present energy delivery system, an energy resource must be continuously available at locations accessible to the population centers. These requirements have led to the almost exclusive exploitation of the fossil-fuel-fired steam-boiler power plant. Eighty per cent of the electricity produced in the United States is generated in this manner in spite of its intrinsic inefficiency and severe pollution problems.

About 7 per cent of our power is generated in hydroelectric plants by the energy of falling water. This is the only significant utilization in our country of the vast resources of energy of motion that exist on the surface of the earth. The atmosphere and the waters of the earth are in constant motion. Waves pound ceaselessly on every shore. Ocean tides drawn by the moon's gravitational force rise in vast rhythmic movements twice every twenty-four hours. Water is drawn up into the atmosphere by the warmth of the sun and falls again as rain, flowing down streambeds, thundering over waterfalls, or pouring in steady powerful currents to the sea. Prevailing winds constantly move the entire atmosphere of the planet from west to east. Sometimes the winds blow in wild gales; sometimes they barely ruffle the surface features of the earth. Here and there, in weak fault zones along the planet's crust, geysers of hot steam erupt, driven by heat energy deep in its molten core.

Taken together, these natural happenings involve enormous amounts of energy in the form of motion. And motion can be made to do useful work with a high degree of efficiency. The technology for converting this type of energy to work is already well developed. Windmills and water wheels were among man's first mechanical inventions for harnessing energy. Propellor-driven generators can convert the wind's energy into electricity at an efficiency of somewhere between 60 and 80 per cent, as compared to the maximum efficiency of 40 per cent for the steam-boiler method. Although wind power is available everywhere in the world, it is extremely variable. In order to use it effectively there must be a system for storing the electric energy in peak hours and using it when the wind has died. The electrolysis of water to produce hydrogen and oxygen would provide such a system and make wind power a practical resource.

Turbines driven by moving water also operate at high efficiencies, and could be used to harness the energy of the tides. There are certain particularly favorable locations where the ocean tides have sufficient range to make them feasible as a power source. In the Bay of Fundy, for example, the average range of the tides is 18 feet. It has been estimated that 100 thousand megawatts of electric power could be harvested from tidal energy in the United States if all the favorable locations were developed. Because these special geographical sites are rarely convenient to heavy-load centers, transmission systems would be needed to deliver the power. Underground transmission would make it possible to harvest this energy without destroying the unique beauty of these deep coastal bays and inlets. A tidal power plant of 240 megawatts has recently been placed in operation by the French government in the estuary of the Rance River, where the tides average 27 feet. But no use has been made of tidal energy in the United States.

Wind power and all forms of water power have the great advantage of not introducing any waste heat or other polluting by-products into the biosphere. They utilize energy sources that are constantly renewed by nature and would, therefore, help to conserve our nonrenewable resources.

"Free" energy to operate electric turbines is also available from underground reservoirs of water heated under pressure to 300-700 degrees Fahrenheit by thermal processes inside the earth. Weaknesses in the earth's crust along major fault lines such as those in California, Japan, Italy, and the oceanic ridges allow this water to erupt into steam geysers with sufficient pressure to run electric generators. Although the steam from these geysers contains some impurities—salt, carbon dioxide, and hydrogen sulphide—these chemicals can be removed, and some can be marketed as by-products. Since there is no combustion, the total pollution produced by geothermal power is less than that of coal-fired power generation.

It is estimated that the United States may hold between 5 and 10 per cent of the world's geothermal reserves. If tapped, these fields could contribute as much as 5 per cent of the total power generated in this country. But the United States has lagged behind other nations in utilizing this resource. The Italians, for instance, recognized its value very early. As far back as 1777 they recovered borax from the mineral-rich steam of fumaroles in Tuscany. Now Italy harvests about 400,000 kw of electricity from geothermal steam. New Zealand, Iceland, Japan, and Russia have also tapped this power source. The U.S. has one geothermal plant (near San Francisco) that has just recently been expanded to produce 192,000 kw of power.

Federally financed geologists are investigating the possibility of an important geothermal field near the Imperial Valley. They are hoping that this will prove to be a very large underground reservoir which would not only provide an enormous amount of power but would also provide millions of gallons of irrigation water for the desert lands of our Southwest. The Mexican government has already developed several geothermal power sites near the California border. It is estimated that a 1000-mile region straddling this border may contain more power capacity than the six highly polluting new coal-burning plants built or planned for the Southwest at the Four Corners.

The most abundant source of free energy for the earth is the sun, which every day pours upon our planet more energy than we could dream of using even in our present power-hungry society. The total energy striking the land area of the United States is about 240,000 kwh per person per day. But this energy is spread very thin; and dilute energy is difficult to convert into work. The energy must be collected from a large area and the equipment is apt to be large and cumbersome. These disadvantages offset the fact that the energy itself is free.

A number of very ingenious devices have been invented, however, for harnessing this energy of the sun. These devices heat water, pump water, cook food, warm houses, and even air condition. In November 1955, a World Symposium on Applied Solar Energy was held in Arizona. Nine hundred engineers, scientists, and businessmen came from countries around the world bringing along an astonishing assortment of interesting inventions such as a solar furnace capable of producing temperatures up to 4000 degrees Centigrade, a solar still for desalinating sea water, and a solar battery that turns sunlight directly into electricity.

In judging the feasibility of these inventions and estimating their usefulness, considerable attention was given to the unit cost of the energy provided by the solar power and comparing this cost with the unit cost of electricity in the location where the utilization would occur. Solar devices could rarely compete on this cost basis with electricity even in 1955. Today with the price of electricity lower relative to the rest of the economy these inventions utilizing solar energy suffer an even greater disadvantage in competitive pricing. However, if the electric rates reflected the real cost of producing the power—if they included the extra expense of avoiding unnecessary damage to the environment—then many of these solar-energy systems would be economically competitive and their development could be commercially pursued on this basis.

Even with these economical handicaps imaginative engineering has created designs that look promising for the utilization of solar energy. Dr. Maria Telkes invented a solar-heating system which has been installed in two experimental houses. Based on this design, the Center for Energy Management and Power at the University of Pennsylvania hopes to develop solar-heating components that can be mass-produced at a price competitive with conventional heating systems. The solar-heating would cause no air pollution and require no transmission of energy.

Several large-scale systems for converting solar energy into electricity have also been proposed. Professors N. C. Ford and J. W. Kane of the University of Massachusetts at Amherst suggest that large areas in the desert could be utilized for collecting solar energy by means of a system of plastic lenses. These lenses, mounted and geared to track the position of the sun, would concentrate the incident energy onto boilers. By this method water could be heated up to 1500 degrees centigrade and used in steam turbines to generate electricity. Such lenses are presently mass-produced and available at a price that would—Ford and Kane believe—make this system economically feasible.

Another proposal for collecting sunlight on desert areas has been made by Dr. Aden Meinel, director of the Optical Sciences Center of the University of Arizona, and his wife, Marjorie Pettit Meinel. Their system for concentrating solar energy envisions large collecting areas covered with selective films. These special materials act like one-way glass, transmitting the sunlight that falls on the top surface and reflecting or absorbing energy that tries to pass back out in the other direction. There are selective films available today capable of concentrating energy 10 to 20 times which cost about one dollar per square meter. The Meinels estimate that a desert area of one square mile could collect by this system enough energy to produce nearly 100 megawatts of electricity. If 14 per cent of the desert regions of the United States were turned into efficient solar farms, one million megawatts of power could be produced, approximately the amount of additional power that may be "needed" in the United States between now and 1990.

A similar plan, using a large recepter area of 50 to 70 square miles covered with solar cells instead of selective film, has been suggested by W. R. Cherry of NASA's Goddard Space Flight Center. Solar cells, which convert the sun's energy directly into electricity, were invented at Bell Laboratories in the 1950's and have been perfected for use in the space program. At the present time the cost of solar cells makes this plan prohibitively expensive, but Mr. Cherry suggests that design development and mass production could reduce the price dramatically.

All these plans are dependent on a good energy storage system, because the energy we receive from the sun varies enormously from hour to hour and from day to day. They also depend on a good energy transmission system to make them economically and aesthetically attractive. Large solar farms would be ideally located in the desert areas of the Southwest, where the sun delivers maximum energy per square foot and the land is relatively valueless for other purposes.

Power produced in any such large-scale projects would have to be transmitted great distances to service cities and industrial areas in other parts of the country.

None of the exotic energy resources described above would be sufficient alone to replace present power production methods and provide for the projected increased demand. But the simultaneous exploitation of several of these resources, added to the generating facilities we already have, would relieve the current energy crisis and at the same time add no extra burden of pollution to the biosphere. Hopefully they might tide us over until the time when fusion power finally becomes a reality.

Atomic energy released by the fusing of two light nuclei appears to be the best hope for meeting future energy needs. Fusion power would be capable of providing almost unlimited power with less pollution per kilowatt than is produced by our present methods. The atoms that would be used for fuel (isotopes of hydrogen and lithium) are widely available and nearly inexhaustible. Because the products of the reaction are not radioactive (with the exception of tritium, which may be reusable as a fusion fuel), there should be no problem of disposing of radioactive wastes. Better thermal efficiency can be achieved with the fusion reaction, resulting in less waste heat than is created by the atomic plants in operation today. However, scientists have not yet been able to harness the fusion process so that it takes place in a controlled and contained manner. They hope that a breakthrough in this field will occur soon; but even after such a hope is realized, fifteen or twenty more years and a great deal more money will probably be required before the process yields safe, reliable electric power.

During these interim years the electric companies plan to increase generating capacity manyfold. If the new facilities are built on the current designs of atomic-fission reactor and coal-fired plants, the earth's resources of fossil fuels, uranium, and thorium will be consumed at a prodigal rate. Air and water pollution will inevitably increase. On the other hand, increased power needs could be met by utilizing some of the earth's "free energy" resources which are non-polluting and constantly renewed by nature.

The electric industry, however, backed by the Atomic Energy Commission, is betting its money on a different interim solution—the liquid-metal-fast-breeder reactor. Breeder plants create more fissionable material than they use up and thereby offer a way of stretching the world's supply of uranium and thorium. Under normal operation they create less radioactive waste and less waste heat than the atomic reactors in current use. The breeder program is especially attractive to the power industry because it promises cost advantages in the production of electricity. It is estimated that the liquid-metal-fast-breeder reactor will produce power at a savings of from .025 to 0.1 cent per kilowatt over present reactors.

However, breeder plants present greater public safety hazards than the type of atomic fission plant now in common use. Breeder plants will contain large amounts of plutonium 239, a highly toxic radioactive element used in making atomic bombs. Plutonium is virtually nonexistent in the earth's natural crust and is very long-lived; its half-life is 24,000 years. Plutonium is known to be carcinogenic and so dangerous to human health that the "permissible" air concentrations are about one part per million billion. If the breeder reactors are developed according to the projections of the Atomic Energy Commission, by 1980 thirty tons of plutonium may be produced annually. The designers expect that elaborate protective devices built into the plants will prevent the release of any of this extremely dangerous element. However, it is not possible to guard against all acts of violence or natural catastrophes—earthquakes, fires, war, or sabotage. Such events could cause the release of an element that would contaminate a large area of the earth for more than 24,000 years.

If breeder plants are built as projected in the 1980's, they will certainly need to be located far from population centers or indeed from any human habitation, for the rural citizen should be able to expect the same protection from dangerous installations as the city resident. These extraordinary siting requirements for breeder plants will cause massive damage to the environment unless a really effective underground transmission system has been developed. The electric industry was spending less than $1 million annually as recently as 1970 on the development of such a system. On the other hand, it is backing the breeder reactor program much more substantially than any other area of research. According to the estimates of the Energy Policy Staff, in 1970 non-federal spending on this program amounted to $25 million. Federal spending that same year was about $100 million.

Many people feel that it is unwise to promote the development of the fast-breeder reactor, which would cause an accumulating burden of a very poisonous element. They believe that this amount of money would be much more wisely spent on speeding the development of fusion power, which offers a safer source of energy. Funding by both government and industry is very inadequate for this important research project. Figures for fusion research in 1970 were $1 million for non-federal and $30 million for federal spending. In 1971 the government increased its support of breeder reactor research to $180 million and at the same time reduced its commitment to fusion power. The Energy Policy Staff estimated that $200 million annually over about 20 years would be needed to bring fusion power to realization.

Improvements on Old Systems

In addition to discovering ways of harnessing alternate energy sources, research and development programs could find more efficient ways of using the old resources and could effect many improvements in present power-plant design. But in these areas also the electric industry is not devoting sufficient money to development. For example, the removal of a large percentage of the sulphur and nitrogen oxides from stack emissions is possible; the sulphur and nitrogen can be marketed as by-products. Combined financing by industry and federal government in 1970 was only about one-tenth the amount needed to fund the development of this technology. In the meantime, large plants are being built with stacks 1000 feet tall in order to distribute the sulphur dioxide over a wider area of countryside. "This," says American Electric Power Company's 1971 *Annual Report*, "remains today the best known means of effective SO_2 dispersal."

Similarly, research on the technology of dry cooling towers to dispose of waste heat is not being pursued in this country, in spite of the fact that this type of tower appears to offer the solution least damaging to the environment. In dry towers the hot water is channeled through tubing exposed to an air flow and gives up its heat to the air without evaporation. Dry towers, therefore, do not add water vapor to the atmosphere; they have the added advantage of providing much greater flexibility in the siting of power plants because a large source of cooling water is not necessary. But dry cooling towers cost more than wet cooling towers and cooling lakes. To date, only one dry tower has been installed in this country, on a relatively small power plant; and this tower was manufactured in Germany.

Many plants being built today will empty their cooling water directly into lakes or rivers. The Commonwealth Edison plant at Zion, Illinois, the Donald C. Cook plant in Michigan, and the Kewaunee and Point Beach plants in Wisconsin are expected to pour 5.5 billion gallons of water warmed 20 degrees Fahrenheit or more into Lake Michigan every day.

A number of interesting ideas for using waste heat constructively have been suggested. Waste heat might be used for heating and cooling greenhouses to permit year-around raising of special crops; to warm city streets in winter and keep them free of ice and snow; or for space heating, especially in new urban developments. Systems of this kind would be effective if peak needs were supplemented by additional heating sources. Systems that combine winter space heating with air conditioning in summer are theoretically possible making use of the waste heat. Another possibility with great potential in desert areas near salt water is the use of waste heat to desalinate sea water. The water could be used for cooling, then desalinated and made available for irrigating the desert.

In most of these proposals flexibility in choice of site is important. Here again the design of a really efficient transmission system must go hand in hand with the exploitation of the various advantages to be gained by appropriate choice of plant location.

It has been suggested that it would be advantageous to locate power plants underground. In the case of nuclear plants, this construction would reduce the hazard of accidental discharge of radioactive or toxic elements. It would also reduce the aesthetic impact and make it possible to locate the plants closer to populous areas, thereby minimizing the need for transmission lines. Only one underground nuclear plant has been built anywhere in the world. Again the innovation has been spearheaded abroad—a joint French-Belgian venture in the populous Meuse Valley.

The discovery of better techniques for producing and delivering power offers our best hope of conserving our environment, while at the same time providing more energy for an expanding economy. By taking the initiative in exploring these fields, the United States could provide technical leadership for the world. Since research and development programs require capital and know-how, the United States is in an ideal position to make this positive contribution toward improving the standard of living of all peoples. This would be a more constructive way of achieving world leadership than being the world's most prodigal consumer of the earth's limited resources.

Who Are the True Luddites?

The power industry accuses conservationists of advocating a return to horse-and-buggy days. Those citizens who question the rapid escalation of *present* technology are branded as "Luddites"—a name given to a group of Englishmen who advocated the destruction of all machinery in the early nineteenth century.

But in failing to invest money and effort in research for an ongoing technology, the barons of the power industry are the true Luddites of today. Clinging to old methods and refusing to fund research for better solutions adequately, they demonstrate a fundamental lack of faith in the ability of scientific innovation to improve our way of life. If they believed in it they would invest in it. "We're not spending money on technology we know won't work," said Donald Cook, exemplifying the industry's attitude toward research.

Instead of improving their technology, they are attempting to force scaled-up monsters of out-dated technology on the American people.

Using the power of monopoly, the power of eminent domain, and the power of their vast financial empire, they are able to overcome almost all protest. And where they encounter unusual resistance they mount a scare campaign of desperate haste.

The psychology of desperate haste should be avoided at all costs. Haste is contrary to the most important conditions of scientific research. Haste is the cause of inadequate testing of side effects; it is the reason for putting into service equipment that has not been adequately checked; it is the reason for clinging to old engineering methods (the new take too long to develop).

Better research and development programs in the electric industry would conserve space and preserve the natural beauty of our countryside. Money now spent on promotion and advertising, diverted to research, would make approximately seven times more financing available for these programs.

Money to fund the needed programs could also be raised by taxation. This system has been initiated in Maryland, where a law passed recently imposes a surcharge on electricity of one tenth of a cent per kilowatt hour. The revenues are being used for research projects and to conduct environmental impact studies. Several spokesmen from both government and industry have proposed a federal tax on electricity. A one per cent tax would bring in approximately $300 million annually for a joint government-utility research fund.

If seven to ten times more money and effort on a national basis were spent on research for better methods of producing and transmitting power, by 1980 we should have solved many crucial problems. According to estimates prepared by the Energy Policy Staff of the Office of Science and Technology, an effort of this magnitude over the next decade can be expected to produce satisfactory technology for the removal of the sulphur, nitrogen oxides, and particulates from stack gases, the constructive use of waste heat, the design of a practical underground transmission system, more efficient power production cycles, and techniques for utilizing some of our free-energy resources. We would be much nearer to harnessing fusion power. Through this research we would learn how to produce more power with less pollution than we have today, and would have a sound scientific foundation on which to base the next stage of expanded energy use. Facilities designed and ordered at that time would incorporate the improved technologies. Only on this kind of real advance can we build an expanding economy that can be truly described as *progress*.

15

Progress Comes to Zilchville, U.S.A.

Two bulldozers from the Ohio Power Company rumbled slowly up the narrow lane to Aunt Kiz's place very early on a summer morning. A thin haze obscured the newly risen sun and dimmed its light, as though nature had drawn a veil over the day. The mist lay like white clouds in the little valleys, softening the outlines of the fence rows and blending the colors of golden wheat field with green pasture and yellow drifts of sweet clover. In the stillness of the morning the bawling of a bereaved cow mourning for her weaned calf could be heard, sad and insistent from far away.

It was known in the village that construction was to start that day, and in spite of the early hour several spectators were standing by the gate watching the progress of the giant bulldozers.

The right-of-way on the Hostler property had been bought and paid for, although Aunt Kiz had fought bitterly against it until the very last. Her two nephews, who owned one-third shares in the place, had supported her resistance for many months. After some difficulty they had found a lawyer who handled the negotiations. The first half-dozen law firms the Hostlers had approached regretted that they were unable to take the case because of conflict of interests. They explained that the electric companies regularly hire law firms on a retainer basis. Usually the best-known and most prestigious firms in each county are in the employ of the electric industry. Even though these firms might not be involved at the moment in any legal proceedings relating to rights-of-way, the fact that they are regularly receiving money from the electric industry means that they are not free to represent private citizens in a case against any electric company. The retainer system gives the electric industry another immediate advantage. Wherever and whenever an argument crops up, they are assured of top legal representation. In addition, of course, there are thousands of lawyers employed full time by the industry.

However, the Hostlers had finally managed to locate a sympathetic attorney and had negotiated for over a year with the power company. The prices offered for the right-of-way slowly rose. Every week or so Aunt Kiz received a long-distance call from one of the nephews, urging her to agree to a settlement. But Aunt Kiz was adamant; she did not want to settle at any price. Finally the Ohio Power Company had filed legal proceedings and the case was due to come up within a month.

"Fine," said Aunt Kiz over the phone. "Let it go to court!"

At that point the nephews came down to Ohio to talk to her. It would be very expensive, they explained, to argue the case in court. Because of the way the laws were written, it was considered impossible to win an appropriation case, and the legal fees might amount to $10,000. Several thousand had already been spent on negotiations. The Hostlers had also been advised that the settlement granted by a court would be considerably less than the sum the power company was offering. All in all, by taking the case to court they would probably lose $15,000 or $20,000. Of course, if Aunt Kiz wanted to squander that amount of money herself—. But Aunt Kiz did not have any money to squander. She was absolutely dependent on the little bit of capital she had put away.

"Why should it cost so much to defend what belongs to me?" she protested.

Protests, however, could not alter the facts. Aunt Kiz had finally given in because she could see no alternative. But she felt that in doing so she was betraying a trust, not only to this family home but, in some more general sense, to the future. As she signed the document she said, "I have never done anything in my life that seemed as wrong to me as signing this paper!"

It was just a few weeks later that the bulldozers and the construction crew came down the lane. They were not deflected for a moment by Taterbug Brown's "bend in the line." They tore down the orchard fence and crashed through the thick grass, toppling the sweet cherry tree loaded with dark red fruit. Then with one crunching shove they demolished the ancient vine-covered springhouse.

For four generations this springhouse had served the Hostler family, providing coolness throughout the summer heat. It had been retired now from active service for almost fifty years. But during all that time the fresh little spring had bubbled up irrepressibly inside the springhouse, run down through the stone troughs, and spilled over into the rivulet that meandered through the calf-lot and eventually joined the larger flow in Slick-a-way Creek. Now two scoops of the bulldozer destroyed this natural drainage and the water began to back up into an ugly yellow puddle. The machines sloshed in and out of the mud, spreading it throughout the construction area. Here and there the crushed forms of ripe cherries made red pockmarks in the yellow mud.

"I know it's foolish to feel sentimental about a springhouse," Aunt Kiz said. "It wasn't useful anymore, but it brought back lots of good memories. It made me think of the times I used to help Mama set the crock of fresh milk in the cold water that stood in the stone trough. Then we skimmed the cream off the crock that had been there all night. Mama had an old tin skimmer with holes in it big enough to put a pencil through and anything that went through those holes wasn't cream." She smiled wryly. "I remember once we had a city friend come to visit and he stood watching the skimming. I can still see his face—sort of amazed and horrified at the same time. Finally he said, 'What's that on top of the milk? Cheese?' Imagine, he didn't even know what real cream looked like!"

We were silent for a while, watching the bulldozers scoop up the stone troughs from the ground and deposit them with other rubble on the little flowerbed by the screen porch. In this shady corner Aunt Kiz had coaxed a number of delicate wildflowers and perennials to grow—yellow primroses and white columbine, bleeding hearts and fragrant lilies-of-the-valley. The flowerbed was buried deep now under the debris. Then clay and brush were added to the pile, branches from the white peach tree still festooned with green fruit, and the knobby trunk of an old apple tree. On top of the pile a robin's nest was impaled on a broken bough. The pink featherless bodies of two baby robins lay limp on the tangle of brush nearby. But one baby bird still clung to the nest, its mouth open, blindly searching the empty air.

Finally the day's work was finished. The construction crew went away and the clash of the machines was stilled. I stayed on for a while with Aunt Kiz and we sat on the porch watching the sunset glow slowly fade on the fields. The acreage beyond the orchard was planted in wheat this year. It was ready for harvesting, its mobile surface stirred by wind ripples. With the first shadow of dusk the fireflies came, just a few sparks of lights at first, then magically multiplying until it seemed as though all the stars in the Milky Way had descended over the field of ripe grain.

By the time another harvest is ripe, I thought, the fireflies will have gone somewhere else. They will never light the sky over this field again. And the robins will not return to nest in the apple trees. Between where I sit and the fields will rise a steel skeleton twelve stories high.

For a long time now Aunt Kiz and I had been making small talk, unable to voice the feelings that were uppermost in our hearts. Finally I said, "Perhaps you should consider moving out. Sell this place and buy a small house in some part of the country that hasn't been spoiled." The words came out wrong and I was sorry the minute I had said them.

Aunt Kiz shook her head. "I can't see a scrap of sense in that," she said angrily. "No other place would ever seem like home to me.

Everyone says, 'Sell out and move on,' and when that place is threatened—what then? . . . 'Sell out and move on.' Soon there will be no beautiful places left because this is happening everywhere—all over America."

"There are always a few crackpots who feel sentimental about dear old grandfather's place," remarked a power company executive, "but we have standard ways of dealing with them. In rural regions and small towns there's rarely any effective opposition. It's easy to force our way through Zilchville."

One of the time-tested strategies that the power companies have found very effective in squeezing out the last remnants of resistance is to proceed with construction as though they owned the entire right-of-way. The bulldozers were already hard at work on the Hostler property although a number of other landowners had not yet signed. Several cases were being vigorously protested. Clovis Strasbow's legal action was still pending. Nevertheless, construction was going full speed ahead. After most of the towers are in place, the neighbors paid, the construction brought up to the boundary lines of the few property owners who are still holding out, most of the remaining opposition crumbles and the owners settle out of court. The few cases that go to court have always been decided in favor of the utility company. How could you reasonably ask a company to move a multi-million-dollar line that is 90 per cent finished?

The law as presently constituted gives almost unlimited power to the electric industry. It allows them to uproot people as casually as they topple the trees with their bulldozers. It allows them to decide arbitrarily that many scenic and fertile parts of our country will be used as utility areas to serve the big cities. By these materialistic and self-serving policies, big industry is progressively narrowing the number of options open to Americans. It is destroying the special character of life in the little villages and on the farms. The small indigenous communities with roots that go back deep into our history are being squeezed out in order to accelerate the building of high-rises and suburban housing. As Spengler predicted half a century ago, "The

Pierson Studio from *Fair Is Our Land,* 1942

The peaceful rural scenes will have been replaced by an industrialized countryside where no one would want to live anymore.

Billy Davis III

giant city sucks the country dry, insatiable and incessantly demanding . . . till it wearies and dies in the midst of an almost uninhabited waste of country."

This wasteland is growing all around us and within us, for the landscape is part of the fabric of ourselves. Like Walt Whitman's child who went forth every day and became part of all he saw, the early lilacs become part of us, and "grass, and white and red morning glories, and the song of the phoebe bird." Now these country scents and sounds are being replaced by plumes of smoke and acid rain. These, too, become part of us. And the whine of electricity passing overhead and the barred shadow of cold steel across the fields. These become part of us.

The wasteland is growing every day. There are very few places left where people can retain a continuity with the generations that bred them and with the landscape that has become part of them. Those who do still know and love this way of life are powerless to protect it against the march of "progress." Because they are few, their rights are trampled. Because they do not have large financial resources, they are unable to fight effectively for these rights in the courts. Multi-billion-dollar combines like the public utilities have the power to force upon people their goal of an all-electric mechanized megalopolis, fed and energized by an industrialized countryside where no one would want to live anymore.

In that world of tomorrow the peaceful rural landscapes will have disappeared, and the little villages like Laurel and Sparksville will have become truly Zilchvilles. Inexorably, step by step, this destruction is proceeding, not because people want it but because big business is insatiable for growth and profits. To feed this ever-increasing appetite, large monopolies such as the electric industry abuse the discretionary powers entrusted to them by the American people. Unless the people recognize what is happening to their land and take power back into their own hands, the destruction will not end until it has created one vast Zilchville, U.S.A.

Epilogue

A year and a half has passed since *Power Over People* went to press, and during those eighteen months "progress" has continued its inexorable march across our countryside. As you drive east now on Ohio Route 180, approaching Laurel, you go over a sharp crest and see three enormous towers straddling the tops of the wooded hills that cradle the town. These stark silhouettes are the dominant feature in this landscape that was once so purely pastoral. Clovis Strasbow and I both resisted for three years the appropriation of the right-of-way for the transmission line across our farms. We finally settled out of court because the laws of Ohio did not allow any issue to be raised in court other than the value of the land appropriated. We did not lie down in front of the bulldozers. That type of simple physical response might make us heroes for a day but we knew it could cause only a brief hesitation in the accelerating rhythm that had searched out the quiet places we wanted to protect. We intend instead, as soon as the line has been energized, to use our land as outdoor laboratories. In cooperation with university scientists we will conduct some of the biological experiments that are needed to understand the long-term effects of these lines on living things.

In the meantime, hundreds of additional miles of extremely high-voltage lines are under construction and rights-of-way are being bought for many more. Plans call for at least 2500 circuit miles of

765-kv lines by 1980. Lines of this voltage have already been built in seven states.

Other confrontations with the energy industry have also been resolved in the last year and a half. Most have been lost; a few have been won. In Belmont County, Ohio, the little community of Barnesville lost its fight to preserve the countryside around the town from the strip-miners. In December 1972 traffic was bypassed around Interstate 70 for about twenty-four hours. A temporary bed of gravel was laid, and on this roadbed the huge Gem of Egypt moved majestically across the interstate highway to start its work of stripping the farmland and forested hill country around Barnesville.

In Illinois another citizen group met with greater success. The farmers near Seneca had fought doggedly for three years to prevent Commonwealth Edison from appropriating about 4500 acres to use as cooling lakes for the huge new nuclear plant at that site. Under pressure to start their building program, the company finally signed an agreement reducing the size of the land taken for the cooling lake to 2100 acres actually needed for the first two generating units. Two additional units planned for the 1980's will require other cooling facilities but land that may be necessary for this purpose must be appropriated when the need actually materializes.

Perhaps the most important event that has occurred in the last eighteen months is the fuel crisis, which has brought dramatically to the attention of people all over the world the finiteness of the earth's fossil fuel resources. It has, of course, been obvious for a long time that we must encounter the limit of these resources sometime in the near future. When we use each ten or twenty years more energy than has been used during the entire history of man up to that time, it is apparent that we must be rapidly approaching the ultimate limit. The surprising thing about the energy crisis is that it occurred earlier than even the most pessimistic theorists had predicted. We are very fortunate that the crisis was precipitated by a sudden change in the distribution of energy resources rather than by the imminent exhaustion of all our reserves. By forcing people to adapt now to lower rates of consumption and focusing on the finiteness of our resources, the energy crisis may have a beneficial effect on the attitudes and policies that were responsible for the prodigal and rapidly accelerating use of energy. If the challenge of this crisis is successfully met we will find ways of reducing waste and utilizing energy more efficiently. We will also mount a greatly expanded research and development program to develop better methods of generating and transmitting power and to discover ways of utilizing resources that are constantly renewed, such as the energy from the sun, the winds, and the tides. The alternative "solution"—more rapid discovery and exploitation of our remaining fossil fuel reserves—will, at best, only postpone the day of reckoning by a few years.

The extremely high-voltage transmission line is a good example of the wasteful design practices that have helped bring about the energy crisis. The 765-kv lines are designed for minimum expense in transmission and consequently these lines waste a considerable amount of electricity. Since electricity is relatively cheap it is less expensive for the power companies to lose this power than to build larger lines that would carry this great amount of electricity without significant loss. The 765-kv lines now operating lose, on the average over a year's time, enough energy in corona discharge alone to provide the electrical needs of forty to fifty homes along *each mile* of right-of-way. In addition there is a loss from heating in the conductors, and when the line is fully loaded the heat losses amount to several times as much as the corona loss. Both of these losses can be reduced by using larger conductors and more conductors to carry the power. An increase of about 25 per cent, for example, in the diameter of the conductors could conserve enough energy along each mile of these lines to provide the electricity for approximately one hundred homes. Along 1000 miles of line the power for a hundred thousand homes is wasted for the sake of "economy." Today lines should be designed for maximum efficiency, and development work should be speeded up on the new techniques such as supercooled dc cables that offer the closest approach to transmission without loss.

Energy wasted in corona discharge also contributes to photochemical oxidant pollution levels that have been found to be surprisingly high in many rural areas. In the late spring of 1973 two papers were published in a professional journal by engineers employed by American Electric Power Company. These papers purport to demonstrate for the industry that no measurable amounts of ozone are produced by 765-kv transmission lines. The reports are actually a restatement of "token experiments" which had been shown to me earlier and which appeared to me to be seriously deficient from a scientific point of view (see above, pp. 151 and 153). The data included the field measurements that had been used to try to persuade the attorney general of Ohio that these lines were safe. The reorganization of this material into a single report makes it appear to be the result of one cohesive and well-planned scientific study, but in fact it is just a collection of hastily conducted efforts to get a quick and favorable answer to the question: how much ozone is produced by 765-kv lines? By this means American Electric Power hoped to lay to rest one of the principal issues raised in *Power Over People.* As the authors say in the introduction to their article: "the studies commissioned by American Electric Power Company or its subsidiaries were aimed at providing a strong technical case to allay environmental fears." Studies that are conducted to prove a point usually fall short of serious scientific research and it is not surprising that the experiments from which these data were drawn violated a number of important scientific principles. The number of readings taken was too small to be significant, the reporting of data was incomplete, important variables were not controlled; in many cases they were not even recorded. None of the tests, for example, specify the voltage being run over the lines when the measurements were made. But other articles published by American Electric Power state that these lines were operating at 8 to 10 per cent below rated voltage. A reduction of 10 per cent in voltage reduces the corona loss and the related phenomenon of ozone production *substantially to zero.* Perhaps the most significant deficiency in these studies was the absence of the questioning attitude which is the very core of all true scientific research. When contradictory and unexpected results were obtained, no attempts were made to discover the reasons for the unexplained phenomena.

The most striking result of the measurements that were reported was the surprisingly high level of oxidant readings that were recorded in several locations. For example, levels as high as .064 ppm (parts per million by volume) were measured in open country in Indiana on a sunny day. According to present theory of air pollution chemistry, sunlight causes a rise in ozone levels at the earth's surface only in the presence of other pollutants such as nitrogen dioxide. But nitrogen dioxide levels were believed to be negligible in these test locations. When such high oxidant readings were found, the nitrogen oxide levels should have been checked. If these levels were consistently very low then the high oxidant readings contradict accepted air pollution theory. This important fact, if properly investigated, might have led to significant new discoveries. No serious scientist would pass up such an opportunity. But this fact was not explored; no questions were asked. This omission is typical of the manner in which these studies were pursued. They were planned and conducted to support a conclusion already taken in advance. They were not real scientific attempts to increase our understanding of these phenomena.

Since high levels of photochemical oxidant have been shown to occur in characteristic country areas in the central and eastern United States, the unnecessary addition of any oxidant to the atmosphere probably constitutes a violation of the ambient air standards. In Ohio, for instance, the maximum one-hour concentration should not exceed .06 ppm. This figure was approached or exceeded in 4 out of the 20 locations tested. To verify whether these lines do indeed violate air standards, it would be necessary to conduct continuous monitoring under the low point of lines in several representative locations during an entire year when the lines are fully energized. If such measurements have been made they have not been made public; and no measurements have been made to detect the presence of hydrogen peroxide, hydroxyl radicals, singlet oxygen, nitrous acid, or other toxic elements created by corona discharge.

Another biological hazard created by these extremely high-voltage lines is emerging as a very significant danger in the light of scientific studies that have been conducted recently. Under these lines and within about 200 feet of the edge of the right-of-way there is a very strong electric field. I had commented on the magnitude of this field (see Chapter 4, p. 47) and expressed my concern about the long-term effects of living most of one's life in such an unnatural environment. But it was not until several months after the book was published that I learned about a very simple and dramatic way to make this field visible. I received a letter from a woman in Oregon who had read *Power Over People;* she enclosed a picture of herself standing under a 500-kv line holding a lighted fluorescent bulb in her bare hand. This bulb was not touching any wire or any metal object. It just lighted up all by itself. My husband and I duplicated the experiment under a 765-kv line. We found that fluorescent bulbs lighted up about 100 feet outside the right-of-way and glowed brighter as we approached the line. Holding the bulbs up in the air, we could feel the vibrations of the alternating current passing through them, making them quiver almost like living things.

In a sense, everything in the vicinity of these high-tension alternating current transmission lines is "plugged into" electricity. There are small but continuous currents running all the time in the ground, the plants, the rocks, the farmer on his tractor. There is a very high voltage difference between one point in space and another point, say a yard away. The potential difference between these two points may be as high as 20,000 volts under a 765-kv line, and the strong electric fields extend way beyond the right-of-way. Many people have their homes, their back yards, the fields where they work for many hours a day in regions where light bulbs will light up in your bare hand.

What does it do to living things to spend a great deal of time in such intense electric fields and to be subjected to electric currents running continuously through them? Considering the importance of this question, one might assume that it had been quite thoroughly investigated by the power companies before building such lines; but a search of the scientific literature revealed the fact that this subject has not been adequately studied. The few research projects that have been done show that there probably are profound effects caused by these fields.

About eight or ten years ago, when American utilities were starting to use extra-high-voltage transmission, two tests were conducted by the companies. In one experiment they exposed 22 male mice *to* strong electric fields for 6½ hours a day over a 10½-month period.

These mice were bred at regular intervals with females that had not been exposed. The number and size of the progeny produced were considered to be a measure of any latent effect which might have been passed on by the exposed fathers. The average size of the male progeny was found to be significantly smaller than normal. On the basis of this finding, the scientists recommended that further studies be conducted, but no follow-up studies have been reported in the literature.

The other study involved ten linemen who did repair work on 345-kv lines. The company studied these men for nine years, doing seven complete medical examinations on them. At the end of the time three of the ten linemen had reduced sperm count. However since sperm count had been quite variable throughout the various medical examinations, the report stated that it would be hazardous to draw any conclusion on this effect from such a small sample. Studies of this type, of course, should examine a much larger number of people over a longer period of time.

A much more thorough examination of this problem, however, has been conducted in Russia. In 1962, after the first Russian 500-kv lines had been operating for several months, men working at the substations began to complain of headaches and a general feeling of malaise. They associated these symptoms with exposure to the electric fields. The Russians made a long-term study of these effects with systematic medical examination of about 250 men working at 500-kv substations. These results were compared with medical examinations of men working at lower-voltage substations. The studies showed that long-time work at 500-kv substations without protective measures resulted in "shattering the dynamic state of the central nervous system, heart and blood-vessel system, and in changing blood structure. Young men complained of reduced sexual poten[cy]." The severity of these effects appeared to depend on the length of stay in the field. As a result of their findings the Russians set up rules for exposure of their personnel to electric fields. According to these regulations no one should be exposed for any time to fields over 25,000 volts per meter without special protective screens or wire cages. In a field of 25,000 v/m the maximum exposure should be 5 minutes. The number of minutes allowed increases as the field strength diminishes. At 10,000 v/m, 180 minutes is allowed, and 5,000 v/m is taken as the level where any length of exposure can be considered safe. It is fortuitous that the field required to light up a 40-watt fluorescent bulb is approximately 6,000 v/m; so the area of space that the Russians would judge to be in a biologically hazardous "field of influence" can be easily mapped out by walking in the vicinity of a transmission line holding a 40-watt fluorescent bulb. At the height of a man's head near the low sag point of our 765-kv lines, this field of influence encompasses a band about 400 feet wide, twice the width of the right-of-way. This space includes homes, yards, roads, and thousands of acres of farmland where people work for many hours a day. A man riding on a tractor under a 765-kv line would be exposed to an electric field so intense that the Russians would not allow their personnel to be exposed to it for more than 5 minutes. In this country we believe that we have more respect for human rights and human life than the Russians have. Yet we have not made as much effort to understand the effects of exposing people to intense electric fields. Nor have we taken comparable measures to protect our citizens from this type of biological damage.

Just recently, the United States Navy has undertaken a series of research projects on the effects of electromagnetic fields very similar to those caused by high-voltage transmission lines. These studies were commissioned in order to answer questions raised about Project Sanguine, the enormous low-frequency antenna that the Navy wants to build in order to communicate with its submarines around the world. As originally proposed, this antenna was to occupy about 21,000 square miles in northern Wisconsin. Lines carrying 14 kv were designed to radiate as elements of an antenna in a large grid pattern. The people of Wisconsin were concerned about the effects that might be caused by this project, particularly the possibility of biological damage from the electromagnetic fields. As a result of the protest the Navy did commission some studies of possible biological effects. The first studies were completed in 1970 and showed a number of effects, ranging from increased mutation rate in fruit flies to high blood pressure in beagle dogs.

The studies, however, were considered to be inconclusive and follow-up experiments were ordered. By this time the Navy had decided to reduce the voltage they would use so that the electromagnetic fields would be less powerful than the ones originally planned. So far only a few of the follow-up studies have been completed. One of these, another mutation study of fruit flies, showed no mutation effects; but the fields used in this study were only one-half as strong as those used in the first study where increased mutation was found, and only one-thousandth as strong as those at ground level under 765-kv lines.

Another study that has been completed showed that some types of cardiac pacemakers are susceptible to disruption even from very small electromagnetic fields. These results are confirmed by other experiments that have been reported in medical journals. Currents as small as 50 microamps disrupt the operation of demand pacemakers, and currents of this size are running all the time in the bodies of people standing on the ground under the 765-kv lines. Just walking under these lines would be extremely hazardous for anyone using one of these pacemakers.

An experiment on nerve and brain tissue demonstrated that electric fields over a certain intensity can affect the function of these very delicate tissues. Brain organelles were found to be especially sensitive to electrical effects. Scientists at Illinois Institute of Technology Research Institute who completed these studies in 1970 suggested that more research is needed to define the exact level at which these disruptive effects begin to occur. Another study just recently completed at the University of California at Los Angeles shows a significant alteration in interresponse time of laboratory animals. This is a very sensitive indication of changes in brain function that may have important long-term effects. These results were obtained by exposing monkeys to fields only one-hundredth as strong as the fields to which people are exposed under 765-kv lines.

The studies conducted in connection with Project Sanguine were commissioned by the Navy, which uses federal funds and, therefore, is required by law to issue Environmental Impact Statements. The environmental report must demonstrate that the installation planned will not be dangerous. Up to the present time the power companies that have built 765-kv lines have not come under any jurisdiction of this kind. They have not been required to give any consideration to whether the thousands of miles of lines they are installing will subject people to hazardous living conditions. Now, however, 765-kv lines are being planned to emanate from nuclear plants and, since the Atomic Energy Commission is a federal agency which must consider the environmental impact resulting from these plants, impact statements are required for the transmission lines. This new situation offers an opportunity for citizens concerned about these lines to demand that the impact statement be not just a superficial whitewash of the problem but be based on a real in-depth scientific study of the long-term effects of extremely high-voltage lines on living things.

During the last year or two the dangers associated with these extra-high-voltage lines have been discussed in the press and other news media. This publicity has resulted in some constructive changes. Better state legislation governing transmission lines has been passed or is pending in several states (see chart, pp. 202-4). Ohio has new legislation which requires certification for transmission lines begun after October 1974. A few states such as New York and Virginia require consideration of environmental impact.

Although the power companies themselves have not officially responded to the issues raised in *Power Over People,* they have been quietly altering the design of their 765-kv lines to reduce the amount of corona loss and even the electric field effects. In 1971 they increased the diameter of their lines from 1.165 inches to 1.386 inches (see above, p. 141). In 1973 Detroit Edison Company announced that their 765-kv lines planned to come from the Greenwood Energy Center would have conductors of 1.425 inches diameter and that they expected to use the standard 40-foot minimum height from ground. A few months later after a protest meeting in Michigan where the hazards resulting from the intense electric field were discussed, Detroit Edison published a statement saying that they would build their lines higher with a minimum distance of 55 feet from ground. This is an important change that will reduce the strength of the electric field at ground. It is apparent that this change was made in response to the questions raised at the protest meeting. Actually, Detroit Edison is especially vulnerable on this point because they intend to run two 765-kv lines in parallel, thereby increasing the fields to which people are subjected in locations between the two lines. Nevertheless, it is encouraging that public exposure of these issues can bring about design changes.

However, it must be remembered that in relation to the total problem these favorable responses are like the flicker of a few candles in the night. Much more public interest and pressure must be brought to bear before there is a really significant alteration in the enormous facilities that the power companies plan to build during the rest of this century. The policy of generating and transmitting power at minimum cost without considering the side effects must be challenged throughout the United States. The results of this policy are jeopardizing hundreds of thousands of people and are dissipating resources that can never be replaced: our dwindling fossil fuel reserves, our prime farmland needed for adequate food production, and most important of all—the beauty of our country. There is a certain finite but still glorious amount of beauty in the world. Unlike our fossil fuel reserves or even our productive land, the total amount of beauty cannot be stretched by learning to use it more efficiently. The loveliness of nature is easy to despoil and impossible to recreate. When the radiant complexion of our land, like a young girl's face, is mutilated by scars, the most heroic efforts and plastic surgery can remove the marks themselves, but the fragile perfection of nature can never be truly recaptured.

Appendix A

WHAT YOU CAN DO ABOUT TRANSMISSION LINES

If a right-of-way agent comes to your door:

1. Ask for the facts. If the agent himself is not able to answer all your questions, tell him that you want this information from the company: What is the maximum voltage that the line will carry? How many circuits will be carried on this right-of-way? How tall will the towers be and where will they be placed? What is the minimum distance from the line to the ground? What size conductors will be used? How much television and radio interference will the line cause? What other properties will it cross? Will the line originate from a nuclear plant?

2. Don't sign any documents that the agent brings with him. Even if you do eventually decide not to fight the installation of this particular line, you should first learn all the facts, consult with your neighbors, and negotiate directly with the company through an attorney. The last people to sign an agreement often end up with the largest payment, sometimes as much as seven times the original offer. You *should* receive a high payment because the line will

have an important effect on the quality of your life and property.

3. Immediately following the agent's departure, telephone neighbors who will be involved. Advise them not to sign and suggest that they call their neighbors. Assemble all concerned people as rapidly as possible because a group has power that an individual lacks.

4. Find a lawyer knowledgeable in environmental law. This is not easy, but lawyers unacquainted with the field may claim that fighting right-of-way appropriations is hopeless. However, in recent cases power companies have been required to justify the need for the line, to consider alternate routes, and to assess the environmental impact.

5. Check just as soon as possible with your state regulatory commission for electric utilities. In many states (see chart, Appendix B) the utility must obtain a license from the commission before a transmission line may be built. Objections to the line should be filed *before* the license has been granted. If possible, obtain a public hearing to consider the public safety and environmental impact of the line. This action is especially important if the line is to carry 765 kv or 500 kv. If the line emanates from a nuclear plant, ask to see the Environmental Report that the company must file with the Atomic Energy Commission. If you do not feel that the safety and environmental hazards have been properly considered in the report, then you and your neighbors should petition to intervene in the AEC licensing procedures.

Even if your property is not directly involved in the right-of-way, you can help protect the safety and attractiveness of your community. You can join with groups of property-owners requesting public hearings or intervening in licensing procedures. One of the most effective avenues for community action is to work for better laws at the state level. Does your state require licensing and public hearings? Are environmental and safety factors carefully considered before licenses are granted? If not, draw these problems to the attention of your state legislators. Organize citizen support for better laws to protect people against dangerous installations that can be imposed upon them under the existing system.

Appendix B

STATE REGULATIONS FOR ELECTRIC TRANSMISSION SYSTEMS, SEPTEMBER 1973

	LICENSE REQUIRED FOR CONSTRUCTION OF TRANSMISSION LINES		PUBLIC HEARINGS HELD		
	In All Instances	*Under Special Circumstances*	*In All Instances*	*If Requested by an Intervenor*	*At the Discretion of the Commission*
Alaska				x	x
Alabama		x			x
Arizona[1]	x Over 115 kv		x Over 115 kv		
Arkansas	x				x
California[1,2]	x Over 200 kv				x
Colorado	x Most instances			x	x
Connecticut[1]	x		x		x
Delaware				x	x
District of Columbia	x All transmission underground				
Florida[1]	x			x	
Georgia					

	LICENSE REQUIRED FOR CONSTRUCTION OF TRANSMISSION LINES		PUBLIC HEARINGS HELD		
	In All Instances	*Under Special Circumstances*	*In All Instances*	*If requested by an Intervenor*	*At the Discretion of the Commission*
Hawaii					x
Idaho[1]	x		x		
Illinois	x		x		
Indiana					
Iowa	x			x	x
Kansas	x			x	x
Kentucky	x				x
Louisiana					
Maine[1]	x		x Over 125 kv		
Maryland[1]	x		x		
Massachusetts	x		x		
Michigan		x			x
Minnesota[3]					
Mississippi	x		x		
Missouri	x Outside certified area		x Outside certified area		
Montana[1]	x				x
Nebraska	x		x Public meetings held		x
Nevada	x			x	x
New Hampshire[1]	x		x		
New Jersey				x	x
New Mexico	x		x		
New York[1]	x Over 125 kv		x		
North Carolina					
North Dakota					
Ohio	x After Oct. 1974		x After Oct. 1974		
Oklahoma[4]					
Oregon	x If condemnation necessary		x If condemnation necessary		

	LICENSE REQUIRED FOR CONSTRUCTION OF TRANSMISSION LINES		PUBLIC HEARINGS HELD		
	In All Instances	*Under Special Circum-stances*	*In All Instances*	*If Requested by an Intervenor*	*At the Discretion of the Commission*
Pennsylvania[2]					x
Rhode Island[4]		x		x	x
South Carolina[1]	x Over 125 kv		x		
South Dakota[3]					
Tennessee	Except TVA	x	Except TVA		
Texas[3]					
Utah	x				x
Vermont[1]	x		x		
Virginia[1]	x			x	x
Washington					
West Virginia[1]	x Over 200 kv			x	x
Wisconsin[2]	x				x
Wyoming	x				x

1. Legislation passed since January 1, 1968.
2. New legislation pending.
3. No state regulatory commission.
4. No answer received from the state commission. Latest information 1968.

Notes and References

The numbers refer to
the corresponding text pages.

Chapter 1

5 *Quotations:* cited by Richard Brewer, "Death by the Plow," *Natural History,* 79, No. 7 (August-September 1970): 28.

Chapter 2

16 *Conductor diameter and corona discharge:* Nestor Kolcio, Vincent Caleca, Stephen J. Marmaroff, and W. L. Gregory, "Radio-Influence and Corona-Loss Aspects of AEP 765kv Lines," *IEEE Transactions of Power Apparatus and Systems,* PAS-88, No. 9 (September 1969); see graphs on p. 1352.

17 *Dubos quotation:* René Dubos, "An Answer? We Don't Even Know the Question," *Chicago Tribune Magazine,* April 12, 1970.

17 *Conductor diameter and voltage:* Kolcio *et al.,* "Radio-Influence and Corona-Loss," Figure 13, p. 1352.

19 *Kelvin's "law":* L. O. Barthold and H. G. Pfeiffer, "High-Voltage Power Transmission," *Scientific American,* 210 (May 1964): 41.

21 *AEP's lines:* Kolcio *et al.,* "Radio-Influence and Corona-Loss," pp. 1353-54.

21 *Plan to go to 1500 or 2000 kv:* see, for instance, "Tests on 765kv Line Enhance Outlook for UHV," *Electrical World,* September 1, 1971, p. 39.

22 *Constituents of the atmosphere:* Theo Loebsack, *Our Atmosphere,* trans. by E. L. and D. Rewald (New York: The New American Library, 1961), Mentor edition, p. 22:

Typical Components (% volume)			*More Variable Components*	
Nitrogen	N_2	78.084	Water vapour	H_2O
Oxygen	O_2	20.946	Ozone	O_3
Argon	A	0.934	Hydrogen peroxide	H_2O_2
Carbon dioxide	CO_2	0.033	Ammonia	NH_3
Neon	Ne	0.00001818	Sulphuretted hydrogen	H_2S
Helium	He	0.00000524	Sulphur dioxide	SO_2
Methane	CH_4	0.000002	Sulphur trioxide	SO_3
Krypton	Kr	0.00000114	Carbon monoxide	CO
Hydrogen	H_2	0.0000005	Radon	R
Nitrous Oxide	N_2O	0.0000005	Dust, soot, salt particles	
Xenon	Xe	0.000000087		

22-23 *Corona discharge:* John A. Coffman and William R. Browne, "Corona Chemistry," *Scientific American,* 212 (June 1965): 90-100.

24 *Ozone characteristics:* "Ozone," *Kirk-Othmer Encyclopedia of Chemical Technology,* 2nd ed., Interscience Publishers, 14 (1967): 410-29.

25 *Ozone in smog:* A. J. Haagen-Smit, "The Control of Air Pollution," *Scientific American,* 210 (January 1964), 25-31.

26 *Singlet oxygen:* "Ozone Reactions Produce Singlet Oxygen," *Chemical and Engineering News,* May 4, 1970, p. 34.

26 *Report on Dr. Khan's work:* "Another Pollution Culprit," *Science News,* December 6, 1969, p. 539.

26 *International conference: ibid.,* p. 538.

27 *Nitrogen dioxide and lung damage: Air Quality Criteria for Nitrogen Oxides,* U.S. Department of Health, Education and Welfare, National Air Pollution Control Administration Publication, No. AP-84, Chapter 9.

27 *Photochemical cycle: Air Quality Criteria for Photochemical Oxidants,* U.S. Department of Health, Education and Welfare, National Air Pollution Control Administration Publication, No. AP-63, p. 2:7.

27 *Nitrous acid and cancer:* Howard J. Sanders, "Chemical Mutagens," *Chemical and Engineering News,* 47 (June 2, 1969): 62-63.

27 *Biological destructiveness of PAN*: Howard J. Sanders, "Chemical Mutagens."

28 *Classifications of radio reception:* Kolcio *et al.*, "Radio-Influence and Corona-Loss," p. 1345, Table II.

28-29 *AEP signal-to-noise estimates: ibid.*, p. 1345. The majority of semi-urban areas receive signals of 60 dB or better. Radio noise expected to be 71 dB at edge of right-of-way in foul weather and 54 dB in fair weather. Signal-to-noise ratio of 16 dB is rated as Class D.

28-29 *Criteria of acceptability: ibid.*, pp. 1345-46.

29-30 *Quotation about voltage:* publicity folder by OVEC-IKEC, *Twins on the Ohio.*

30 *Increase in radio noise and corona discharge with voltage:* Kolcio *et al.*, "Radio-Influence and Corona-Loss," Figure 6, p. 1347, and Figures 12 and 13, p. 1352. See also p. 1344 ". . . for a 10 per cent change in maximum conductor gradient the line RI will change by 6 dB."

30 *Plan to increase voltage on AEP 765-kv lines:* Gregory S. Vassell and Raymond M. Maliszewski, "AEP 765 kv System: System Planning Considerations," *IEEE Transactions on Power Apparatus and Systems,* PAS-88, No. 9 (September 1969): 1328.

30 *Statements about TV interference*: C. F. Clark and M. O. Loftness, "Some Observations of Foul Weather EHV Television Interference," *IEEE Transactions,* 1970, Paper No. 70 TP 104-PWR. See also "Radio Noise Design Guide for High-Voltage Transmission Lines," IEEE Radio Noise Subcommittee Report—Working Group No. 3, *IEEE Transactions on Power Apparatus and Systems,* PAS-90, No. 2 (March-April 1971): 842: "Regarding TVI, the presence of significant amounts of noise generation from the conductor during foul weather has been debated for some time; it is now agreed that for a combination of low signal strengths, high conductor gradients, and foul weather, the generation is, in fact, significant."

30 *Quotation concerning complaints*: Clark and Loftness, "Some Observations of Foul Weather EHV Television Interference," p. 3.

31 *Quotations concerning interference:* Kolcio *et al.*, "Radio-Influence and Corona-Loss," pp. 1355-56.

31 *Canadian engineers' comments: ibid.*, p. 1354.

31 *Quoted statement:* J. Reichman in personal conversation.

31 *Audible noise measurements:* "Radio Noise Design Guide for

High-Voltage Transmission Lines," IEEE Radio Noise Subcommittee Report—Working Group No. 3, *IEEE Transactions on Power Apparatus and Systems,* PAS-90, No. 2 (March-April 1971): 839. Also Gerhard W. Juette and Luciano E. Zaffanella, "Radio Noise, Audible Noise, and Corona Loss of EHV and UHV Transmission Lines Under Rain: Predetermination Based on Cage Tests," *IEEE Transactions on Power Apparatus and Systems,* PAS-89, No. 6 (July-August 1970): 1171-72. Audible noise measurements under lines carrying 745 kv recorded a noise level of 56 dB (A scale) 75 feet from the outer phase: H. C. Barnes, "Preliminary Analysis of Extensive Switching Surge Testing of American Electric Power's First 765 kv Line and Stations," *IEEE Transactions on Power Apparatus and Systems,* PAS-90, No. 2 (March-April 1971): 787.

32 *Description of audible noise levels:* "Industry's Quiet Rush to Silence," *Iron Age,* December 16, 1971, pp. 73-78.

32 *Chicago ordinance on noise levels:* cited in *Chicago Tribune,* July 6, 1972, Section 1-A, p. 1.

32 *Audible noise from power lines;* Juette and Zaffanella, "Radio Noise, Audible Noise," p. 1171 and Figure 5.

32 *Acceptable level for audible noise:* "Tests on 765kv Line Enhance Outlook for UHV," *Electrical World,* September 1, 1971, p. 39.

33 *"The present basic philosophy . . .":* Kolcio *et al.,* "Radio-Influence and Corona-Loss," p. 1343.

33 *Electrical engineers reporting on radio-noise criteria:* "Radio Noise Design Guide," p. 842.

34 *Miles of transmission lines to be built this century: Environmental Criteria for Electric Transmission Systems,* U.S. Department of the Interior and U.S. Department of Agriculture (Washington, D.C.: 1970), p. iii.

Chapter 4

41 *Federal jurisdiction:* the Federal Power Commission also has jurisdiction under Section 20(b) of the Federal Power Act over wholesale rates for electric energy sold in interstate commerce. *United States Law Week,* 40 LW 4141 (January 11, 1972). The Atomic Energy Commission has jurisdiction over nuclear power plants.

41-42 *Interpretation of law of eminent domain:* see Ohio Revised Statutes, Section 4933.15, Section 163.08, and Section 163.09. Although, under the revised code enacted in 1966, the question of necessity may also be raised, the cases decided before 1966 have retained their precedential value; and these cases were decided on the principle that the only issue in question was the value of the property appropriated.

42 *Legal commentary: American Jurisprudence,* Vol. 18, Sec. 108, p. 735.

43 *Capital wealth of electric industry: The Price of Power,* Report of the Council on Economic Priorities (1972), p. 4.

43 *Quotation from Business Week:* cited in *Conservation Foundation Letter,* March 1970, pp. 56-57.

44 *State regulatory commissions: Electric Power and the Environment,* a report sponsored by the Energy Policy Staff, Office of Science and Technology, S. David Freeman, Director (August 1970): 56-57. See also chart on pp. 202-4.

45 *National Electric Code:* Handbook 81 was published in 1961. Very minor revisions were published in 1965 and 1968.

46 *Safe "let-go" thresholds:* Charles F. Dalziel and W. R. Lee, "Lethal Electric Currents," *IEEE Spectrum,* February 1969, p. 46. Also D. F. Shankle, "RI, Audible Noise, and Electrostatic Induction Problems," *1970 Utility Engineering Conference,* March 15-27, 1970, Subject No. 57, p. 5.

46 *Dalziel quotation:* Dalziel and Lee, "Lethal Electric Currents," p. 50.

46 *Magnitude of electrostatic shock currents under 765-kv lines:* Shankle, "RI, Audible Noise," Figure 9, p. 16. See also "Electrostatic Effects of Overhead Transmission Lines," Report of the Working Group on Electrostatic Effects of Transmission Lines, April 15, 1971, *IEEE Transactions,* Paper No. 71 TP 644-PWR, p. 4: "This indicates that lethal currents can be built up on long insulated fences under such [EHV] lines."

46 *Shocks causing involuntary movement:* Charles F. Dalziel, "Electric Shock Hazard," *IEEE Spectrum,* 6, No. 2 (1972): 43. In 1970 the American National Standards Institute adopted the standard of 0.5 milliamps as the maximum allowable leakage current for two-wire portable electric devices. This adoption followed a three-year research project into the danger of reaction currents.

47 *On parking of vehicles:* "Electrostatic Effects of Overhead Transmission Lines," p. 3.

47 *Pasteur's work on electric fields:* René Dubos, *Pasteur and Mod-*

ern Science (Garden City, N.Y.: Doubleday, 1960), pp. 36-37.

48 *Statement on National Electric Safety Code:* personal letter from William J. Meese, Chairman ANSI Standards Committee C-2 on National Electric Code, National Bureau of Standards, National Department of Commerce, May 5, 1970.

49 *Height of AEP lines:* A. James Samuelson, R. L. Retallack, and R. A. Kravitz, "AEP 765kv Line Design," *IEEE Transactions on Power Apparatus and Systems,* PAS-88, No. 9 (September 1969): 1366.

49 *Comment questioning this interpretation of the code:* discussion comment by R. E. Moran following article by Samuelson *et al.,* "AEP 765kv Line Design," p. 1371, including following chart:

	AEP (feet)	NESC (feet)	Margin
Ground clearance	40	37 [pedestrians only]	3
Over roads	45	42	3
Over railroads	50	50	0

49 *AEP engineers' answer: ibid.,* p. 1371.

49-50 *Canadian standards: Canadian Electric Code, Part III.*

50 *Ohio Ambient Air Standards for Photochemical Oxidants:* 1. The maximum one (1) hour arithmetic mean concentration shall not exceed one hundred and nineteen (119) micrograms per cubic meter (0.06 parts per million by volume). 2. The maximum four (4) hour arithmetic mean concentration not to be exceeded more than one (1) consecutive four (4) hour period per year shall be seventy-nine (79) micrograms per cubic meter (0.04 parts per million by volume). 3. The maximum twenty-four (24) hour arithmetic mean concentration not to be exceeded more than one (1) day per year shall be forty (40) micrograms per cubic meter (0.02 parts per million by volume).

51 *Licensing procedures:* bills now pending in several states would require review and licensing of all electrical installations. (See pp. 202-4 for later information, September 1973.)

Chapter 6

61 *Mixed mesophytic forests:* E. Lucy Braun, *Deciduous Forests of Eastern North America* (Philadelphia: Blakiston, 1950), pp. 24, 35, and map inside back cover.

62 *Ohio and mixed mesophytic forest region: ibid.,* p. 35.

63-64 *Guidelines: Environmental Criteria for Electric Transmission Systems,* U.S. Department of the Interior, U.S. Department of Agriculture (1970), pp. 3-26.

66 *Studies at Brookhaven National Laboratory:* George M. Woodwell, "Energy Cycle of the Biosphere," *Scientific American,* 223, No. 3 (September 1970): 70.

67 *Figures on carbon dioxide concentrations: ibid.,* p. 73.

67 *Carbon dioxide and the earth's temperature:* Eugene K. Peterson, "The Atmosphere: A Clouded Horizon," *Environment,* 12, No. 3 (April 1970): 34-35, 39. See also Bert Bolin, "The Carbon Cycle," *Scientific American,* 223 (September 1970): 128-32.

68 *Decline in earth's mean temperature:* Woodwell, "Energy Cycle of the Biosphere," p. 73.

68-69 *Sears quotation:* Paul B. Sears, *Where There Is Life* (New York: Dell, 1962), pp. 214, 197.

Chapter 8

73 *Carson quotation:* Rachel Carson, *Silent Spring* (Boston: Houghton Mifflin, 1962), p. 13.

73 *"The amounts are minute and rapidly dissipated":* statement made by John Tillinghast, Executive Vice President, American Electric Power Company, in personal correspondence, July 14, 1970.

74 *Oxidant injury to vegetation: Air Quality Criteria for Photochemical Oxidants,* U.S. Department of Health, Education and Welfare, National Air Pollution Control Administration Publication No. AP-63, pp. 6:8, 6:9; also George D. Clayton *et al.,* "Community Air Quality Guides, Ozone (photochemical oxidant)," *American Industrial Hygiene Association Journal,* 29 (May-June 1968): 301.

75 *Injury from PAN: Air Quality Criteria for Photochemical Oxidants,* pp. 6:6 to 6:10, 10:3.

75 *Plant sensitivity to ozone: ibid.,* pp. 6:13 to 6:17.

75 *Decline in crop yields: ibid.,* p. 6:22.

76 *Studies at Yale University on rate of photosynthesis:* "Effect of Ozone on Photosynthesis," *Bulletin of the Ecological Society of America,* 51, No. 2 (June 1970).

76 *Measurements of oxidant levels: Mount Storm, West Virginia; Gorman, Maryland; and Luke, Maryland–Keyser, West Virginia, Air Pollution Abatement Activity,* U.S. Environmental Protection Agency, April Pre-Conference Investigations (1971), pp. 1:61 to 1:63.

77 *Results of ozone exposure on laboratory animals:* Clayton *et al.,*

"Community Air Quality Guides," p. 300.

77 *"Mortality is enhanced . . .": Air Quality Criteria for Photochemical Oxidants,* p. 8:33.

78 *Experiments on humans: Air Quality Criteria for Photochemical Oxidants,* pp. 10:5 to 10:8.

78 *Quibbletown incident:* as reported by Donald Jackson, "The Cloud Comes to Quibbletown," *Life,* December 10, 1971.

79 *Ozone as a bactericide: Air Quality Criteria for Photochemical Oxidants,* pp. 6:12, 6:17.

79 *Standards for precipitators:* G. S. Castle, Ion I. Inculet, and K. Irvin Burgess, "Ozone Generation in Positive Corona Electrostatic Precipitators," *IEEE Transactions on Industry and General Applications,* IGA-5, No. 4 (July-August 1969): 489. The electrostatic precipitators designed for cleaning the particulates from stack gases are so designed that they cause more ozone formation than do precipitators designed for recirculating systems. Engineers have not been concerned with the reduction of ozone formation in these installations and, therefore, the total oxidant produced by the electric fields may exceed ambient air standards. It would be ironic if electrostatic precipitators removed particulates at the expense of adding important concentrations of oxidant to industrial effluents.

80 *Reduced fertility in plants and animals: Air Quality Criteria for Photochemical Oxidants,* pp. 6:4, 8:33, and 10:7.

82 *Statement of Committee for Community Air Quality:* Clayton *et al.,* "Community Air Quality Guides," p. 301.

83 *Zamenhof's work:* Howard J. Sanders, "Chemical Mutagens," *Chemical and Engineering News,* 47 (June 2, 1969): 63.

83 *Buffalo and British studies:* R. E. Waller, "Air Pollution and Lung Cancer," *UNESCO Courier,* 23 (May 1970): 30-32. See also *Air Quality Criteria for Particulate Matter,* U.S. Department of Health, Education and Welfare (1969), No. AP-49, p. 170.

83-84 *Rapid aging due to oxidants: Air Quality Criteria for Photochemical Oxidants,* pp. 6:3, 8:9.

84 *Stokinger's experiments:* H. E. Stokinger, *Archives of Environmental Health* (1965), as cited in *Air Quality Criteria for Photochemical Oxidants,* p. 8:10.

84 *Biological effects of free radicals (aging):* William A. Pryor, "Free Radicals in Biological Systems," *Scientific American,* 223 (August 1970): 81.

85-86 *Distribution of oxidants in Los Angeles area: Air Quality Criteria for Photochemical Oxidants,* pp. 3:15 and 3:16.

87 *High Air Pollution Potential Episodes:* Virginia Brodine, "Episode 104," *Environment,* 13, No. 1 (January-February 1971): 3-27.

90-91 *Cesium 137 and strontium 90 concentrations in Lapps and Eskimos:* Joel Alan Snow and Alvin W. Wolfe, "Radioactivity in Arctic Peoples," *Scientist and Citizen,* September-October 1964, pp. 26-33.

91 *Biological concentration of mercury:* Peter and Katherine Montague, "Mercury, How Much Are We Eating?" *Saturday Review,* February 6, 1971, p. 52.

92 *Report on sampling by Bureau of Sport Fisheries and Wildlife: ibid.*

93-94 *Statements about tobacco industry and congressional hearings:* Harold S. Diehle, M.D., *Tobacco and Your Health* (New York: McGraw-Hill, 1969), pp. 112, 118.

94 *Statements about chemical industry:* Frank Graham, Jr., *Since Silent Spring* (Boston: Houghton Mifflin, 1970).

94 *Research and development expenditures: 1970 National Power Survey,* Federal Power Commission, Chapter 22, p. 3: "The total research and development expenditures reported by electric utilities in 1969 was about $40 million, or less than one-fourth of one percent of gross electric operating revenues." Although some increases have been made since 1969, the amounts are still estimated to be considerably under one-third of one per cent. See also statement by Joseph C. Swidler, Chairman of New York State Public Service Commission, at a meeting of the Institute of Electrical and Electronics Engineers, New York City, February 2, 1971. Reprinted in *Congressional Record,* August 3, 1971, Vol. 117, No. 124, S12926-28.

94 *Comment by Nassikas:* quoted in *New York Times,* January 15, 1971, 25:7.

95 *Testing for presence of ozone:* W. B. Kouwenhoven, O. R. Langworth, M. L. Singewalk, and G. G. Knickerbocker, "Medical Evaluation of Man Working in AC Electric Fields," *IEEE Transactions on Power Apparatus and Systems,* PAS-86, No. 4 (April 1967): 506-11.

95 *Quoted description of experiment: ibid.,* p. 508.

96 *Power company official quoted:* John Tillinghast, American Electric Power Company, in personal letter (July 1970).

Chapter 9

106 *Carbon dioxide figures:* Bert Bolin, "The Carbon Cycle," *Scientific American,* 223 (September 1970): 131.

106 *Sulphur dioxide, nitrogen oxides, and hydrocarbons:* Fred Singer, "Human Energy Production as a Process in the Biosphere," *Scientific American,* 223 (September 1970): 186-88.

106 *Particulates in the air:* Henry A. Schroeder, "Metals in the Air," *Environment,* 13, No. 8 (October 1971): 21. Also "Mercury in the Air," Staff Report, *Environment,* 13, No. 4 (May 1971): 24-33.

106 *1968 figure:* U.S. Department of Health, Education and Welfare, "Nationwide Inventory of Air Pollutant Emissions, 1968" (August 1970), Publication No. AP-73, pp. 3, 17.

107 *Power plants on Lake Michigan: Newsletter,* Lake Michigan Federation (February 29, 1972), p. 5.

107 *Heat from power generation by year 2000:* "The Space Available," A Report from the Committee for Environmental Information, *Environment,* 12, No. 3 (March 1970): 4.

107 *Thermal pollution:* John R. Clark, "Thermal Pollution and Aquatic Life," *Scientific American,* 220 (March 1969): 19-27. Also John R. Clark, "Heat Pollution," *National Parks Magazine,* December 1969.

108 *Cooling towers:* Riley D. Woodson, "Cooling Towers," *Scientific American* 224 (May 1971): 70-78. Dry cooling towers do not make use of evaporation. In these towers the heat is removed from the water by circulating the hot water through a closed system of tubes exposed to the air. Such towers are more expensive than wet cooling towers.

110 *Pollution from plants at Four Corners described:* John Neary, "Hello Energy—Goodbye, Big Sky," *Life,* April 16, 1971, p. 64.

110-11 *Projections for new generating capacity in Ohio, West Virginia area: 1970 National Power Survey,* Federal Power Commission, Part II, 2-16, and 2-133. See also Lenard B. Young, regional engineer, Federal Power Commission, *Historic and Projected Electrical Power Requirements of Areas Bordering on Lake Michigan,* address presented at Workshop on Power Siting Problems, 1975-1990 (June 12, 1971), Chicago.

111 *Exemption from sulphur-dioxide standards:* Virginia Brodine, "Episode 104," *Environment,* 13, No. 1 (January-February 1971): 22.

111 *Air pollution measurements in West Virginia and Maryland: Mount Storm, West Virginia; Gorman, Maryland; and Luke, Maryland–Keyser, West Virginia, Air Pollution Abatement Activity,* U.S. Environmental Protection Agency, April Pre-Conference Investigations (1971), pp. 1:1, 1:56, 1:61, 1:62.

111 *Air pollution damage in Pennsylvania:* T. Craig Weidensaul and Norman L. Lacrosse, "Results of a Statewide Survey of Air Pollution Damage to Vegetation," presented at the 63rd Annual Meeting of the Air Pollution Control Association, June 1970.

111 *Acid precipitation:* Gene E. Likens, F. Herbert Bormann, and Noye M. Johnson, "Acid Rain," *Environment,* 14, No. 2 (March 1972): 33-40.

113 *Strip-mining:* Harry M. Caudill, "Lament for the Appalachian Hills," *Junior League Magazine* (November-December 1969): 8-11.

114 *Governor Gilligan quoted:* address to the joint session of the Ohio Legislature, February 29, 1972, as reported in the *New York Times,* March 5, 1972. Stricter strip-mining laws were passed in Ohio in 1972, in spite of the opposition described by the Governor. Hanna Coal Company had threatened to go out of business if the laws were passed but subsequently decided that it could stay in business, after all.

114 *Amount of reclamation for stripped mines:* Caudill, "Lament for the Appalachian Hills," pp. 61-62. Also E. A. Nephew, "Healing Wounds," *Environment,* 14 (January-February 1972): 14.

114 *Schoolteacher quoted:* Ben A. Franklin, "Strip-Mining Boom Leaves Wasteland in Its Wake" *New York Times,* December 15, 1970, 1:1.

115 *Trace metals washed into strip-pits:* Wayne Davis, "The Strip Mining of America," *Sierra Club Bulletin,* 56, No. 2 (February 1971): 4-7.

115 *Mercury in coal beds:* "Coal Fields Fuel Environmental War," *The Plain Dealer* (Cleveland), February 7, 1971.

116 *Wayne L. Hays quoted:* Franklin, "Strip-Mining Boom."

116 *Telephone operator quoted:* "Coal Fields Fuel Environmental War."

116 *Ford Sampson quoted:* Franklin, "Strip-Mining Boom."

117 *Hendrysburg and Egypt:* "Coal Fields Fuel Environmental War." See also Richard C. Widman and William D. McCann, "Strip Mine Blasting Angers Valley; 40 File Damage Claims," *The Plain Dealer* (Cleveland), February 8, 1971.

117 *GEM of Egypt:* Franklin, "Strip-Mining Boom."

120 *American Electric Power Company's housing development: Annual Report,* 1970, p. 28.

123 *R. W. Hatch quoted:* Richard C. Widman and William D. McCann, "Strip Mine Industry Facing Attacks," *The Plain Dealer* (Cleveland), February 11, 1971.

124 *Electric rates:* Commonwealth Edison's residential rates, 1972.

124 *Efficiency of home furnaces versus electric heating:* Claude M. Summers, "The Conversion of Energy," *Scientific American,* 224 (September 1971): 162.

125 *Advertisement:* Edison Electric Institute.

125 *Use of fossil fuels for power generation:* Federal Power Commission, 1970 *National Power Survey,* Chapter XI, Table XI-6.

125-26 *Pollution caused by electric heating:* Dean E. Abrahamson, *Environmental Cost of Electric Power,* A Scientists' Institute for Public Information Workbook, 1970, p. 30.

126 *Demand of an all-electric building:* "Underground Power Transmission," A Report to the Federal Power Commission by the Advisory Committee on Underground Transmission (April 1966): p. 5.

126 *Quotation about electric heating: ibid.,* pp. 30-31.

126 *Bulk electric rates:* cited in "Another Lost Frontier," *Forbes,* August 15, 1972, pp. 29-30.

126 *Demand of a steel mill:* "Underground Power Transmission," p. 5.

126 *Electric demand in primary metals industry:* Abrahamson, *Environmental Cost of Electric Power,* p. 30.

127 *Growth of aluminum industry:* Barry Commoner, Michael Corr, and Paul J. Stamler, "The Causes of Pollution," *Environment,* 13, No. 3 (April 1971): 7.

128 *19 million electrically heated homes by 1980:* Abrahamson, *Environmental Cost of Electric Power,* p. 30.

129 *Present power use compared to 1950: ibid.,* p. 3.

129 *Projected power use:* these estimates were based on an annual increase of 9 per cent, the most recent growth rate as reported in *Electric Power and the Environment,* Energy Policy Staff, Office of Science and Technology, S. David Freeman, Director (August 1970), p. 2. These estimates were corrected downwards for a smaller population growth as indicated by latest birthrate trends. The industry's estimates were reported in Abrahamson, *Environmental Cost of Electric Power,* p. 4.

131 *Gas used to generate power: 1970 National Power Survey*, Federal Power Commission, Chapter XI, Table XI-2.

131 *New York City EPA Study:* reported in *Newsweek*, May 3, 1971, p. 82.

131 *Donald Cook quoted: Industry Week*, September 21, 1970, p. 48.

132 *John W. Simpson quoted: Sierra Club Bulletin* (March 1971): 11.

133 *White House expert quoted: Newsweek*, May 3, 1971, p. 82.

133 *Surveys of operating delays: Letter of the Conservation Foundation*, March 1970, p. 11. See also 1970 Federal Power Survey, Chapter 16, as cited in *Letter of the Conservation Foundation*, June 1972, p. 5. Atomic Energy Commission Chairman James R. Schlesinger said that delays in nuclear plants had been due "only in small measure to the activities of environment groups." (In testimony before the House Merchant Marine and Fisheries subcommittee on wildlife conservation, March 22, 1972.)

133-34 *Carl Bagge quoted:* "Electrical Power Famine to Hit U.S.," *Environmental Action*, June 25, 1970, p. 5.

Chapter 11

140 *Ohio Civil Rule* 33: This is the interrogatory rule and is based on proposed Federal Rule 33. "Any party may serve any other party written interrogatories to be answered by the party served. . . . Each interrogatory shall be answered separately and fully in writing under oath, unless objected to, in which event the reasons for objection shall be stated in lieu of an answer."

141 *Corona discharge as function of conductor diameter:* Nestor Kolcio, Vincent Caleca, Stephen J. Marmaroff, and W. L. Gregory, "Radio-Influence and Corona-Loss Aspects of AEP 765kv Lines," *IEEE Transactions of Power Apparatus and Systems*, PAS-88, No. 9 (September 1969): graph on p. 1349.

141 *Quotation from AEP engineers: ibid.*, p. 1355.

Chapter 12

144 *Power use in India:* Dean E. Abrahamson, *Environmental Cost of Electric Power*, A Scientists' Institute for Public Information Workbook, 1970, pp. 3, 4. See also *Statistical Abstract of the United States*, 1967, U.S. Department of Commerce, pp. 863, 878. The average per capita figures represent each individual's share of the total amount of power consumed in that country.

145 *Legend as told by Gamow:* George Gamow, *One, Two, Three . . . Infinity* (New York: The New American Library, 1947), Mentor edition, p. 19.

145-46 *Doubling of power plants:* "The Space Available," Report from the Committee for Environmental Information, *Environment,* 12, No. 2 (March 1970): 4.

147 *Pipeline engineer quoted:* Charles G. Siegfried, "Multiple Use of Rights-of-Way for Pipe-Lines," *Pipe Line News,* August 1971, pp. 8-15.

147 *Recommendation for use of rights-of-way: Environmental Criteria for Electric Transmission Systems,* U.S. Department of the Interior, U.S. Department of Agriculture (1970), p. 27.

148 *H. C. Barnes quoted:* "Tests on 765kv Line Enhance Outlook for UHV," *Electrical World,* September 1, 1971, p. 39.

148 *Estimate of miles of new line: Environmental Criteria for Electric Transmission Systems,* p. iii.

149 *Acres needed for cooling lakes:* John R. Clark, "Thermal Pollution and Aquatic Life," *Scientific American,* 220 (March 1969): 24.

149 *New cooling tower:* General Electric Company has developed a new cooling tower shaped like a stadium. Their studies show that the plume of hot moist air from these new towers penetrates clouds and inversion layers more efficiently than the plumes from conventional types of cooling towers. They also estimate that the costs will be competitive with conventional towers. As reported by Gene Smith in *New York Times,* April 7, 1971, p. 57.

Chapter 13

151 *Field measurements:* as reported to author by American Electric Power Company, August 14, 1971.

152 *Impressions of attorney general's representative:* as reported to author by Barry Smith in personal interview, May 1972.

152-53 *Attorney General's statement:* News Release, Office of the Attorney General, William J. Brown, Columbus, Ohio, March 8, 1972.

153 *U.S. Forest Service proposal:* as reported to author by Leon S. Dochinger in personal interview, September 1971.

153 *Three-day field study:* Battelle Memorial Institute, "Oxidant Measurements in the Vicinity of 765 kilovolt Power Lines." As reported December 3, 1971, to American Electric Power Service Corporation.

154 *"The questions raised by these lines . . .":* letter from Attorney General William J. Brown to Kenneth E. McIntyre, Colonel, Corps of Engineers, April 5, 1972.

Chapter 14

155 *Research by electric companies:* Federal Power Commission statement cited in "Electrical Power Famine to Hit U.S.," *Environmental Action,* June 25, 1970, p. 5. See also Irvin L. White, "Energy Policy-Making," *Bulletin of the Atomic Scientists,* October 1971, p. 23.

157 *Information on oil-insulated cables:* "Underground Power Transmission," A Report of the Federal Power Commission by the Commission's Advisory Committee on Underground Transmission (April 1966): pp. 14, 23.

158 *Cost figures for underground distribution: ibid.,* p. 6.

159 *Dc underground transmission:* P. H. Rose, "Underground Power Transmission," *Science,* 170, No. 3955 (October 16, 1970): 271. See also Lawrence Lessing, "DC Power's Big Comeback," *Fortune,* September 1965, p. 74.

159 *Sodium cable shelved:* as reported in *Chemical and Engineering News,* August 10, 1970, p. 15.

160 *I.B.M. study:* Donald P. Snowden, "Superconductors for Power Transmission," *Scientific American,* 226 (April 1972): 91.

160 *Superconducting dc lines:* R. W. Meyerhoff, "Superconducting Power Transmission," *Cryogenics,* April 1971, p. 99. See also Lawrence Lessing, "New Ways with Less Pollution," *Fortune,* November 1970, p. 79.

160-61 *On $8-million pilot program:* Lessing, "New Ways with Less Pollution," p. 80. See also News Release, Union Carbide, June 11, 1969.

161 *On $2-million program:* News Release, Edison Electric Institute, November 12, 1971. The $2.11 million program will be funded in the amount of $1.69 million by EEI and $422,000 by the U.S. Department of the Interior.

161 *Feasibility of superconducting transmission:* Meyerhoff, "Superconducting Power Transmission," p. 93.

161 *General Electric cryogenic cable:* as reported by Gene Smith, "G.E. Discloses Transmission Gain," New York Times, June 1, 1972. See also *Wall Street Journal,* June 1, 1972, p. 15.

161-62 *SF_6-insulated cables:* as reported in *Business Week,* March 20, 1971, p. 58: ITE Imperial Corporation of Philadelphia is building a substation for Consolidated Edison at a Brooklyn power plant and also a short transmission line near Peekskill, N.Y. Cleveland Electric will have a short transmission line (SF_6-insulated) at its Eastlake plant. ITE has utilized transmission technology acquired from Sprecher and Schuh (a European company). Detroit Edison has contracted with a foreign company, Coginel, for a substation to be installed in 1973. See also Rose, "Underground Power Transmission," p. 269.

163 *Information about coal-gas:* Arthur M. Squires, "Clean Power from Coal," *Science,* 169, No. 3948 (August 28, 1970): 821-27. For report on Hygas process see Gene Smith," "A New Fuel: Coal," *New York Times,* February 27, 1972. See also "Current Power Generation Technology," by J. L. Shapiro, *Problems of Electrical Power Production in the Southwest,* hearings before the U.S. Senate Committee on Interior and Insular Affairs, May 24, 1971, p. 542. The Federal Power Commission estimated that pipeline quality could be made from coal for as little as 40 cents per million Btu, compared to the 35 cents per million Btu price for natural gas: *1970 National Power Survey,* II-4:16.

163-64 *Gas made from organic wastes:* Hinrich L. Bohn, "A Clean New Gas," *Environment,* 13, No. 10 (December 1971).

164 *Cost estimates for power plants using coal-gas:* Smith, "A New Fuel: Coal."

164 *Expense of gas transmission versus electric:* Derek P. Gregory, "A New Concept in Energy Transmission," *Public Utilities Fortnightly,* February 3, 1972, pp. 26-27 and Table 2.

165 *Government funding sought:* Gene Smith, "A New Fuel: Coal," p. 3.

165 *Hydrogen lines in Germany:* Christian Isting, "Pipelines Now Play Important Role in Petrochemical Transport," *World Petroleum,* April 1970, p. 41.

166 *Hydrogen system:* Gregory, "A New Concept in Energy Transmission," pp. 21-30. Dr. Gregory estimates the capital costs of 400-kv dc overhead line and hydrogen delivery systems to be roughly comparable (see p. 28, Figure 7). His hydrogen transmission estimate is only 2 to 4 cents per million Btu per hundred miles, and the fixed cost of converting electricity to hydrogen is 37 cents per million Btu. These figures compare well with electric transmission costs if the distance is greater than 500 miles (see Figure 6). See also "Hydrogen: Candidate for Universal Fuel," *Chemical and Engineering News,* April 17, 1972, pp. 34-35.

167 *Proposal of New York Public Service Commission:* as reported by Peter Khiss, "Buried Utility Lines Throughout the State," *New York Times,* February 28, 1971. On January 23, 1972, the *New York Times* reported that this plan was formally recommended to the Public Service Commission by a State Public Service Commission examiner.

167 *Federal Power Commission estimates:* "Underground Power Transmission," p. 41.

169 *Wind power:* Claude M. Summers, "The Conversion of Energy," *Scientific American,* 224 (September 1971): p. 157.

170 *Power from tidal energy: ibid.,* p. 157.

170-71 *Power from geothermal energy: Newsweek,* June 7, 1971, p. 97. See also Robert W. Rex, "Geothermal Energy, the Neglected Energy Option," *Bulletin of the Atomic Scientists,* 27, No. 8 (October 1971): 52-56. Rex estimates U.S. geothermal potential much higher than 5 per cent of our total power production. He cites estimates as high as 10 million megawatts.

171 *Geothermal fields on U.S.-Mexican border:* John Lear, "Clean Power from Inside the Earth," *Saturday Review,* December 5, 1970, p. 59.

171 *Solar energy striking earth's surface:* Farrington Daniels, "The Sun's Energy," *Proceedings of World Symposium on Applied Solar Energy,* Phoenix, Arizona, November 1955 (Stanford Research Institute, 1956), p. 21.

172 *Solar-heated homes:* Maria Telkes, Institute of Energy Conservation, University of Delaware, personal correspondence, July 1972.

172-73 *Collecting solar energy with plastic lenses:* Norman C. Ford and Joseph W. Kane, "Solar Power," *Bulletin of the Atomic Scientists,* 27, No. 8 (October 1971): 27-31. Ford and Kane also suggest that water heated to 1500° C by their lens collecting system could be converted to hydrogen and oxygen by thermal dissociation.

173 *Collecting solar energy with selective films:* Summers, "The Conversion of Energy," p. 158. See also Aden Baker Meinel and Marjorie Pettit Meinel, "Is It Time for a New Look at Solar Energy?" *Bulletin of the Atomic Scientists,* 27, No. 8 (October 1971): 33-37.

173 *W. R. Cherry's suggestion:* "Chance for Solar Energy Conversion," *Chemical and Engineering News,* December 20, 1971, p. 39.

173 *Solar cells:* G. L. Pearson, "Electricity from the Sun," *Proceed-*

ings of World Symposium on Applied Solar Energy*, Phoenix, Arizona (Stanford Research Institute, 1956), p. 281.

174 *Fusion power:* R. F. Post, "Fusion Power, The Uncertain Certainty," *Bulletin of the Atomic Scientists,* 27, No. 8 (October 1971): 42-48.

175 *Breeder reactors:* Glenn T. Seaborg and Justin L. Bloom, "Fast Breeder Reactors," *Scientific American,* 223 (November 1970): 21.

175 *Amounts of plutonium produced:* Donald P. Geesaman, "Plutonium and the Energy Decision," *Bulletin of the Atomic Scientists,* 27 (September 1971): 35.

176 *Research expenditures: Electric Power and the Environment,* a report sponsored by the Energy Policy Staff, Office of Science and Technology, S. David Freeman, Director (August 1970): 44-45.

177 *Research on dry cooling towers: ibid.,* p. 36.

177 *Thermal discharges into Lake Michigan: Power Production and Protection of the Lake,* Proceedings of 2nd Annual Four-State Lake Michigan Conference (Zion, Ill., 1970), p. 107 (Fact Sheet).

177 *Waste heat to desalinate sea water:* Meinel and Meinel, "A New Look at Solar Energy?" pp. 36-37.

178 *Underground plants:* Franklyn C. Rogers, "Underground Nuclear Power Plants," *Bulletin of the Atomic Scientists,* 27 (October 1971): 38-42.

178 *French-Belgian venture:* Harlan Draeger, "Electric Power Industry Going Underground," *Chicago Daily News,* April 26, 1961.

178 *Donald Cook quoted: Forbes,* May 1, 1972, p. 55.

179 *Federal tax on electricity:* proposed by Senator Warren Magnuson of Washington in "Federal Power Research and Development Act," submitted to Congress August 1971.

179 *Estimates of Energy Policy Staff: Electric Power and the Environment,* pp. 44-45.

Chapter 15

187-88 *Spengler quotation:* Oswald Spengler, *The Decline of the West* (New York: Alfred Knopf, 1926), p. 102.

188 *Whitman quotation:* Walt Whitman, "There Was a Child Went Forth," *Leaves of Grass* (New York: W. W. Norton, 1965), p. 364.

Epilogue: 1974

192 *Papers published by American Electric Power Company engineers:*
Harold N. Scherer, Jr., Brendan J. Ware, and C. H. Shih, "Gaseous Effluents Due to EHV Transmission Line Corona," *IEEE Transactions on Power Apparatus and Systems,* PAS-92, No. 3 (May-June 1973): 1043-49.
Maurycy Frydman, Arthur Levy, and Salo E. Miller, "Oxidant Measurements in the Vicinity of Energized 765kv Lines," *IEEE Transactions on Power Apparatus and Systems,* PAS-92, No. 3 (May-June, 1973): 1141-48.

192 *Quotation from AEP article:*
Scherer *et al.*, p. 1043.

192 *Criticism of data reported in AEP papers:*
Thirty-one field measurements of oxidants were mentioned in the AEP papers. Of this number 6 were not reported at all. Of the remaining 25 only 5 were made in rain or snow when significant amounts of corona could have occurred. Only one of these five recorded the vertical distance from line to instrument at the time the measurement was made. This distance was 65 feet, not the 40 feet minimum which occurs four times a mile.

192 *Article stating that lines were operating below rated voltage:*
Raymond M. Maliszewski, Gregory S. Vassell, and Norman B. Johnsen, "Experience with the AEP 765-kv System: Part 11—System Performance, *IEEE Transactions on Power Apparatus and Systems,* PAS-92 (July-August 1973): pp. 1337-47. See especially p. 1941: "It has become AEP's practice to operate each new 765-kv circuit initially at its lowest possible voltage level [689 kv]. The level is then increased as experience is gained with the operation of the new equipment." This practice serves to reduce very substantially during the first test years the effects associated with corona discharge, radio and TV interference, and shock hazard.

193 *Maximum ambient air standards approached or exceeded:*
See data reported by Frydman *et al.*, pp. 1147 and 1148, locations 1, 12, 18, and 20.

194 *Report of two tests conducted by the companies:*
W. B. Kouwenhoven, O. R. Langworth, M. L. Singewalk, and G. G. Knickerbocker, "Medical Evaluation of Man Working in AC Electric Fields," *IEEE Transactions on Power Apparatus and Systems,* PAS-86, No. 4 (April 1967): 506-11. M. L. Singewald, O. R. Langworthy, and W. B. Kouwenhoven, "Medical Follow-up Study of High Voltage Linemen Working in AC Elec-

tric Fields," *IEEE Transactions on Power Apparatus and Systems,* PAS-92 (July-August 1973): pp. 1307-09.

195 *Russian studies on electric fields under power lines:*
V. P. Korobkova, Yu. A. Morozov, M. D. Stolarov and Yu. A. Yakub, "Influence of the Electric Field in 500 and 750 kv Switchyards on Maintenance Staff and Means for its Protection," CIGRE, International Conference on Large High Tension Electric Systems, Paris, August-September 1972. Available from United States Department of Interior, Engineering and Research Center, Washington, D.C.

196 *Report of first Sanguine studies:*
William B. Coate *et al.,* Hazelton Laboratories, Inc., "Project Sanguine, Biological Effects Test Program, Pilot Studies, Final Report," November 1970, prepared for the Department of the Navy, Naval Electronics Systems Command Headquarters, Washington, D.C.

197 *Report of second mutation study for Sanguine:*
"Sanguine System Biological/Ecological Research Program: Summary Status Report," Department of the Navy, Electronic Systems Command, April 1973, p. 3. Also "Electrostatic Effects of Overhead Transmission Lines, Part I: Hazards and Effects," Report of the Working Group on Electrostatic Effects of Transmission Lines, IEEE Transactions Paper, 1971: 71 TP 644-PWR, p. 5 for figures on voltage gradients under 765-kv lines.

197 *Sanguine study on cardiac pacemakers:*
A. R. Valentino, D. A. Miller, and J. E. Bridges, "Susceptibility of Cardiac Pacemakers to 60Hz Magnetic Fields," IIT Research Institute Report. Available from National Technical Information Service, Washington, D.C., as Report #737237.

197 *Experiments on pacemakers reported in medical journals:*
Seymour Furman, Bryan Parker, Martin Krauthamer, and Doris J. W. Escher, "The Influence of Electromagnetic Environment on the Performance of Artificial Cardiac Pacemakers;" *The Annals of Thoracic Surgery,* Vol. 6, No. 1, July 1968: pp. 90-96.

197 *Report of experiment on nerve and brain tissue:*
W. H. Riesen *et al.,* "A Pilot Study of the Interaction of Extremely Low Frequency Electromagnetic Fields with Brain Organelles," Technical Memorandum #3, IIT Research Institute Project E6185, U.S. Naval Electronics Systems Command. Available from National Technical Information Service, Washington, D.C.

197 *Effect of electric fields on interresponse time:*
As reported to the author in a telephone interview by Dr. Ross Adey, University of California at Los Angeles, School of Medicine.

Bibliography

Aaronson, Terri. "Mercury in the Environment." *Environment,* 13, No. 4 (May 1971), pp. 16-28.

Abrahamson, Dean E. *Environmental Cost of Electric Power.* Scientist's Institute for Public Information Workbook. New York, 1970.

Air Quality Criteria for Nitrogen Oxides. Environmental Protection Agency Air Pollution Control Office, No. AP-84. Washington, D.C., 1971.

Air Quality Criteria for Photochemical Oxidants. U.S. Department of Health, Education, and Welfare, No. AP-63, Washington, D.C., 1970.

Air Quality Criteria for Sulphur Oxides. U.S. Department of Health, Education and Welfare, No. AP-50, Washington, D.C., 1969.

Alston, L. L., ed. *High-Voltage Technology.* London: Oxford University Press, 1968.

Anthrop, Donald F. "Environmental Side Effects of Energy Production." *Bulletin of the Atomic Scientists,* 26, No. 8 (October 1970), pp. 39-41.

Arnold, J. R., and Martell, E. A. "The Circulation of Isotopes." *Scientific American,* 201, No. 3 (September 1959), pp. 84-94.

Barnes, H. C. "Preliminary Analysis of Extensive Switching Surge Testing of American Electric Power's First 765kv Line and Stations." *IEEE Transactions on Power Apparatus and Systems,* PAS-90, No. 2 (March-April 1971), pp. 785-99.

Barnes, Howard C., and Nagel, Theodore J. "AEP 765-kv System: General Background Relating to Its Development." *IEEE Transac-*

tions on Power Apparatus and Systems, PAS-88, No. 9 (September 1969), pp. 1313-19.

Barthold, L. O., and Pfeiffer, H. G. "High-Voltage Power Transmission." *Scientific American,* 210, No. 5 (May 1964), pp. 39-47.

Beadle, George W. "Ionizing Radiation and the Citizen." *Scientific American,* 201, No. 3 (September 1959), pp. 219-32.

Benedict, Manson. "Electric Power from Nuclear Fission." *Bulletin of the Atomic Scientists,* 27, No. 7 (September 1971), p. 8.

Bohn, Hinrich L. "A Clean New Gas." *Environment,* 13, No. 10 (December 1971), pp. 4-9.

Bolin, Bert. "The Carbon Cycle." *Scientific American,* 223, No. 3 (September 1970), pp. 125-32.

Braun, E. Lucy. *Deciduous Forests of Eastern North America.* Philadelphia: The Blakiston Company, 1950.

Brewer, Richard. "Death by the Plow." *Natural History,* 79, No. 7, (August-September 1970), p. 28.

Brodine, Virginia. "Episode 104." *Environment,* 13, No. 1 (January-February 1971), pp. 3-24.

Broecker, Wallace S. "Enough Air." *Environment,* 223, No. 3 (September 1970), pp. 27-31.

Bryson, Reid A. "Inadvertent Climatic Modification." *UIR Research Newsletter,* The University of Wisconsin Industry Research Program, February 1967, p. 11.

Castle, G. S. Peter; Inculet, Ion I.; and Burgess, K. Irwin. "Ozone Generation in Positive Corona Electrostatic Precipitators." *IEEE Transactions on Industry and General Applications,* IGA-5, No. 4 (July-August 1969), pp. 489-96.

Charoy, M. A., and Jocteur, R. F. "Very High Tension Cables with Extruded Polyethylene Insulation." *IEEE Transactions on Power Apparatus and Systems,* PAS-90, No. 2 (March-April 1971), pp. 777-85.

Clark, C. F., and Loftness, M. O. "Some Observations of Foul Weather EHV Television Interference." *Transactions of IEEE,* 1970, Paper No. 70, TP104-PWR.

Clark, John R. "Heat Pollution." *National Parks Magazine,* December 1969.

———. "Thermal Pollution and Aquatic Life." *Scientific American,* 220, No. 3 (March 1969), pp. 19-27.

Clayton, George D., *et al.* "Community Air Quality Guides, Ozone (photochemical oxidants)." *American Industrial Hygiene Association Journal,* 29 (May-June 1968), pp. 299-303.

Cloud, Preston, and Gibor, Aharon. "The Oxygen Cycle." *Scientific American,* 223, No. 3 (September 1970), pp. 110-24.

Coffman, John A., and Browne, William R. "Corona Chemistry." *Scientific American,* 212 (June 1965), pp. 90-100.

Committee for Environmental Information. "The Space Available." *Environment,* 12, No. 2 (March 1970), pp. 2-9.

Commoner, Barry; Corr, Michael; and Stamler, Paul J. "The Causes of Pollution." *Environment,* 13, No. 3 (April 1971), pp. 2-19.

Cook, Earl. "The Flow of Energy in an Industrial Society." *Scientific American,* 224, No. 3 (September 1971), pp. 134-48.

Craig, Roy. "Cloud on the Desert." *Environment,* 13, No. 6 (July-August 1971), pp. 20-35.

Curtis, Richard, and Hogan, Elizabeth. *The Perils of the Peaceful Atom.* New York: Doubleday & Company, 1969.

Dalziel, Charles F. "Electric Shock Hazard." *IEEE Spectrum,* 6, No. 2 (February 1972), pp. 41-50.

Dalziel, Charles F., and Lee, W. R. "Lethal Electric Currents." *IEEE Spectrum,* 6, No. 2 (February 1969), pp. 44-50.

Delwiche, C. C. "The Nitrogen Cycle." *Scientific American,* 223, No. 3 (September 1970), pp. 136-48.

Dochinger, L. S., and Bender, F. W. "Chlorotic Dwarf of Eastern White Pine Caused by Ozone and Sulphur Dioxide Interaction." *Nature,* 225, No. 5231 (January 31, 1970), p. 476.

Dochinger, Leon S., and Seliskar, Carl E. "Air Pollution and the Chlorotic Dwarf Disease of Eastern White Pine." *Forest Science,* 16, No. 1 (March 1970), pp. 46-55.

Dubos, René. "An Answer? We Don't Even Know the Question." *Psychology Today,* March 1970.

Ehrlich, Paul R., and Holdren, John P. "The Heat Barrier." *Saturday Review,* April 3, 1971, p. 61.

Eipper, Alfred W. "Pollution Problems, Resource Policy, and the Scientist." *Science,* 169, No. 3940 (July 3, 1970), pp. 11-15.

Eipper, Alfred W.; Carlson, C. A.; and Hamilton, L. S. "Impacts of Nuclear Power Plants on the Environment." *The Living Wilderness,* Autumn 1970, pp. 5-12.

Electric Power and the Environment. A Report Sponsored by the Energy Policy Staff, Office of Science and Technology. S. David Freeman, Director, August 1970.

"Electrical Power Famine to Hit U.S." *Environmental Action,* June 25, 1970.

"Electrostatic Effects of Overhead Transmission Lines." Report of the

IEEE Working Group on Electrostatic Effects of Transmission Lines. *IEEE Transactions* Paper No. 71, TP644-PWR, April 15, 1971.

Eliassen, Rolf. "Power Generation and the Environment." *Bulletin of the Atomic Scientists,* 27, No. 7 (September 1971), pp. 37-42.

Environmental Criteria for Electric Transmission Systems. United States Department of the Interior and United States Department of Agriculture. Washington, D.C., 1970.

Environmental Guidelines. Western Systems Coordinating Council, Robert N. Coe, Chairman, December 3, 1971.

Federal Power Commission. *The 1970 National Power Survey.* Washington, D.C., 1970.

———. *The 1971 National Power Survey.* Washington, D.C., 1971.

———. *Typical Electric Bills, 1970.* FPC R-76. Washington, D.C., December 1970.

Foote, Christopher S. "Photosensitized Oxygenations and the Role of Singlet Oxygen." *Journal of the American Chemical Society,* 1 (April 1968), pp. 104-10.

Freeman, S. David. "Toward a Policy of Energy Conservation." *Bulletin of the Atomic Scientists,* 27, No. 8 (October 1971), pp. 8-12.

Gates, David M. "The Flow of Energy in the Biosphere." *Scientific American,* 224, No. 3 (September 1971), pp. 88-104.

Geesaman, Donald P. "Plutonium and the Energy Decision." *Bulletin of the Atomic Scientists,* 27, No. 7 (September 1971), pp. 33-37.

Gofman, John W. "Nuclear Power and Ecocide: An Adversary View of New Technology." *Bulletin of the Atomic Scientists,* 27, No. 7 (September 1971), pp. 28-32.

Gough, William C., and Eastlund, Bernard J. "The Prospects of Fusion Power." *Scientific American,* 224, No. 2 (February 1971), pp. 50-68.

Gregory, Derek P. "A New Concept in Energy Transmission." *Public Utilities Fortnightly,* February 3, 1972, pp. 21-30.

Gross, Edward. "Another Pollution Culprit." *Science News,* 96 (December 6, 1969), pp. 538-40.

Haagen-Smit, A. J. "The Control of Air Pollution." *Scientific American,* 210, No. 1 (January 1964), pp. 25-31.

———. "Man and His Home." *The Living Wilderness,* Summer 1970, pp. 38-46.

Hasler, Arthur D., and Ingersoll, Bruce. "Dwindling Lakes." *Natural History,* 77, No. 9 (November 1968), p. 8.

Hauspurg, Arthur; Caleca, Vincent; and Schlonnann, Robert. "765-kv

Transmission Line Insulation: Testing Program." *IEEE Transactions on Power Apparatus and Systems,* PAS-88, No. 9 (September 1969), pp. 1356-65.

Hauspurg, Arthur, *et al.* "Overvoltages on the AEP 765-kv System." *IEEE Transactions on Power Apparatus and Systems,* PAS-88, No. 9 (September 1969), pp. 1329-37.

Heggestad, H. E. "Consideration of Air Quality Standards for Vegetation with Respect to Ozone." *Journal of the Air Pollution Control Association,* 19, No. 6 (June 1969), pp. 424-26.

Hendricks, Russel H., and Larsen, Lee. "An Evaluation of Selected Methods of Collection and Analysis of Low Concentrations of Ozone." *American Industrial Hygiene Association Journal,* January-February 1966, pp. 80-84.

Hollaender, A., and Stapleton, G. E. "Radiation and the Cell." *Scientific American,* 201, No. 3 (September 1959), pp. 94-177.

Hore, T., *et al.* "Ozone Exposure and Intelligence Tests." *Archives of Environmental Health,* 17 (July 1968), pp. 77-79.

Hubbert, M. King. "The Energy Resources of the Earth." *Scientific American,* 224, No. 3 (September 1971), pp. 60-88.

Hutchinson, G. Evelyn. "The Biosphere." *Scientific American,* 223, No. 3 (September 1970), pp. 44-53.

"Hydrogen Fuel Use Calls for New Source." *Chemical and Engineering News,* July 3, 1972, pp. 16-18.

"Hydrogen: Likely Fuel of the Future." *Chemical and Engineering News,* June 26, 1972, pp. 14-17.

Inglis, David Rittenhouse. "Nuclear Energy and the Malthusian Dilemma." *Bulletin of the Atomic Scientists,* 27, No. 2 (February 1971), pp. 14-18.

"Is Man Changing the Earth's Climate?" *Chemical and Engineering News,* August 16, 1971.

Jaffe, L. S. "The Biological Effects of Ozone on Man and Animals." *American Industrial Hygiene Association Journal,* 28 (May-June 1967), pp. 267-77.

———. "Photochemical Air Pollutants and Their Effects on Men and Animals, II: Adverse Effects." *Archives of Environmental Health,* 16 (February 16, 1968), pp. 241-55.

Juette, Gerhard W., and Zaffanella, Luciano E. "Radio Noise, Audible Noise, and Corona Loss of EHV and UHV Transmission Lines Under Rain: Predetermination Based on Cage Tests." *IEEE Transactions on Power Apparatus and Systems,* PAS-89, No. 6 (July-August 1970), pp. 1168-78.

Katz, Milton. "Decision-Making in the Production of Power." *Scientific American,* 224, No. 3 (September 1971), pp. 191-200.

Knickerbocker, G. G.; Kouwenhoven, W. B.; and Barnes, H. C. "Exposure of Mice to a Strong AC Electric Field." *IEEE Transactions on Power Apparatus and Systems,* PAS-86, No. 4 (April 1967), pp. 498-505.

Kolcio, Nestor; Caleca, Vincent; Marmaroff, Stephen; and Gregory, W. L. "Radio-Influence and Corona-Loss Aspects of AEP 765-kv Lines." *IEEE Transactions on Power Apparatus and Systems,* PAS-88, No. 9 (September 1969), pp. 1343-55.

Kouwenhoven, W. B.; Langworth, O. R.; Singewalk, M. L.; and Knickerbocker, G. G. "Medical Evaluation of Man Working in AC Electric Fields." *IEEE Transactions on Power Apparatus and Systems,* PAS-86, No. 4 (April 1967), pp. 506-11.

Lagarias, J. S. "Discharge Electrodes and Electrostatic Precipitators." *Journal of the Air Pollution Control Association,* 10, No. 4 (August 1960), pp. 271-74.

Lapp, Ralph E. "The Four Big Fears About Nuclear Power." *The New York Times Magazine,* February 7, 1971, pp. 16-35.

———. "Where Will We Get the Energy?" *The New Republic,* July 11, 1970, pp. 17-21.

Larsen, Lee B., and Hendricks, Russel H. "An Evaluation of Certain Direct Reading Devices for the Determination of Ozone." *American Industrial Hygiene Association Journal,* November-December 1969, pp. 620-23.

Lawrence, R. F. "Transmission in the 70's." *Westinghouse Engineer,* November 1970, pp. 181-84.

Lederberg, Joshua. "Squaring An Infinite Circle." *Bulletin of the Atomic Scientists,* 27, No. 7 (September 1971), pp. 43-45.

Lessing, Lawrence. "DC Power's Big Comeback." *Fortune,* 72, No. 3, (September 1965), p. 74.

———. "New Ways to More Power With Less Pollution." *Fortune,* November 1970, pp. 78-81.

Levins, Philip L., *et al.* "Research Solutions to Pollution." *Industrial Research,* October 1970, pp. P 1-23.

Likens, Gene E.; Bormann, F. Herbert; and Johnson, Noye M. "Acid Rain." *Environment,* 14, No. 2 (March 1972), pp. 33-40.

Lindop, Patricia J., and Rotblat, J. "Radiation Pollution of the Environment." *Bulletin of the Atomic Scientists,* 27, No. 7 (September 1971), p. 17.

Loeb, Leonard B. *Electrical Coronas: Their Basic Physical Mechanisms.*

Berkeley and Los Angeles: University of California Press, 1965.

Loebsack, Theo. *Our Atmosphere*. Translated by E. L. and D. Rewald. New York: The New American Library, 1961.

Loutit, John F. "Ionizing Radiation and the Whole Animal." *Scientific American,* 201, No. 3 (September 1959), pp. 117-38.

McEwan, Murray J., and Phillips, Leon F. "Chemistry in the Upper Atmosphere." *Accounts of Chemical Research,* 3 (January 1970), pp. 9-17.

Main, Jeremy. "A Peak Load of Trouble for the Utilities." *Fortune,* November 1969, p. 116.

"Mercury in the Air." Staff Report, *Environment,* 13, No. 4 (May 1971), pp. 28-38.

Meyerhoff, R. W. "Superconducting Power Transmission." *Cryogenics,* April 1971, pp. 91-101.

Miller, Charles J., Jr. "The Measurement of Electric Fields in Live Line Working." *IEEE Transactions on Power Apparatus and Systems,* PAS-86, No. 4 (April 1967), pp. 493-505.

Montague, Peter and Katherine. "Mercury: How Much Are We Eating?" *Saturday Review,* February 6, 1971, pp. 50-55.

Mount Storm, West Virginia–Gorman, Maryland, and Luke, Maryland–Keyser, West Virginia, Air Pollution Abatement Activity. U.S. Environmental Protection Agency, Research Triangle Park, N.C., April 1971.

Nash, Hugh. "Black Mesa: The Suicide of the Southwest." *Not Man Apart,* 1, No. 1 (December 1970), pp. 12-16.

Nephew, E. A. "Healing Wounds." *Environment,* 14, No. 1 (January-February 1972), pp. 12-21.

Norwood, W. D.; Wisehart, D. E.; Earl, C. A.; Adley, F. E.; and Anderson, D. E. "Nitrogen Dioxide Poisoning Due to Metal-Cutting with Oxyacetylene Torch." *Journal of Occupational Medicine,* 8, No. 6 (June 1966), pp. 301-6.

Oort, Abraham H. "The Energy Cycle of the Earth." *Scientific American,* 223, No. 3 (September 1970), pp. 54-64.

O'Sullivan, Dermot. "Air Pollution." *Chemical and Engineering News,* 48, No. 24 (June 8, 1970), pp. 38-58.

"Ozone." *Kirk-Othmer Encyclopedia of Chemical Technology,* 2nd ed., Vol. 14, Interscience Publishers, 1967, pp. 410-29.

Ozone Chemistry and Technology. Advances in Chemistry Series, Vol. 21. American Chemical Society, Washington D.C., 1959.

Peterson, Eugene K. "The Atmosphere: A Clouded Horizon." *Environment,* 12, No. 3 (April 1970), pp. 32-40.

Photochemical Oxidants and Air Pollution: An Annoted Bibliography. United States Environmental Protection Agency, No. AP-88, 1971.

Platzman, Robert L. "What Is Ionizing Radiation?" *Scientific American,* 201, No. 3 (September 1959), pp. 74-84.

Post, R. F. "Fusion Power: The Uncertain Certainty." *Bulletin of the Atomic Scientists,* 27, No. 8 (October 1971), pp. 42-48.

The Price of Power: Electric Utilities and the Environment. Report of Council on Economic Priorities, New York, 1972.

Proceedings of World Symposium on Applied Solar Energy. Stanford Research Institute, Menlo Park, California, 1956.

Pryor, William A. "Free Radicals in Biological Systems." *Scientific American,* 223, No. 2 (August 1970), pp. 70-83.

"Radio Noise Design Guide for High-Voltage Transmission Lines." IEEE Radio Noise Subcommittee Report—Working Group No. 3. *IEEE Transactions on Power Apparatus and Systems,* PAS-90, No. 2 (March-April 1971), pp. 833-42.

Rex, Robert W. "Geothermal Energy, The Neglected Energy Option." *Bulletin of the Atomic Scientists,* 27, No. 8 (October 1971), pp. 52-56.

Rich, Saul, and Tomlinson, Harley. "How Plants Trap Ozone." *Frontiers of Plant Science,* 22, No. 2 (1970), pp. 4-5.

Rogers, Franklyn C. "Underground Nuclear Power Plants." *Bulletin of the Atomic Scientists,* 27, No. 8 (October 1971), pp. 38-42.

Rose, P. H. "Underground Transmission." *Science,* 170, No. 3955 (October 16, 1970), pp. 267-72.

Samuelson, A. James; Retallack, R. L.; and Kravitz, R. A. "AEP 765-kv Line Design." *IEEE Transactions on Power Apparatus and Systems,* PAS-88, No. 9 (September 1969), pp. 1367-71.

Sanders, Howard J. "Chemical Mutagens." *Chemical and Engineering News,* June 2, 1969, p. 59-67.

Schaefer, Vincent. "The Threat of the Unseen." *Saturday Review,* February 6, 1971, pp. 55-57.

Scheibla, Shirley. "Prometheus Bound." *Barron's,* March 8, 1971, p. 5.

Schroeder, Henry A. "Metals in the Air." *Environment,* 13, No. 8 (October 1971), p. 18.

Schurr, Sam. "Energy." *Scientific American,* 209, No. 3 (September 1963), pp. 110-28.

Seaborg, Glenn T. "The Atoms Are Coming! The Atoms Are Coming!" *Industry Week,* July 5, 1971, pp. 55-58.

———. "On Misunderstanding the Atom." *Bulletin of the Atomic Scientists,* 27, No. 7 (September 1971), pp. 46-53.

Seaborg, Glenn T., and Bloom, Justin L. "Fast Breeder Reactors." *Scientific American,* 223, No. 5 (November 1970), pp. 13-21.

Shankle, D. F. "RI, Audible Noise, and Electrostatic Induction Problems." *1970 Electric Utility Engineering Conference,* Vol. 7, Subject No. 57, pp. 1-17.

Siegfried, Charles G. "Multiple Use of Rights-of-Way for Pipelines." *Pipe Line News,* August 1971, pp. 8-15.

Sierra Club, Midwest Regional Conservation Committee Power Production Task Force. *Report.* February 1971.

Singer, S. Fred. "Human Energy Production as a Process in the Biosphere." *Scientific American,* 223, No. 3 (September 1970), pp. 174-94.

Slappey, Sterling G. "Heading Off the Energy Crisis." *Nation's Business,* July 1971, pp. 26-36.

Snowden, Donald P. "Superconductors for Power Transmission." *Scientific American,* 226, No. 4 (April 1972), pp. 84-91.

"The Space Available." Report from the Committee for Environmental Information. *Environment,* 12, No. 2 (March 1970), p. 4.

Spinrad, Bernard I. "America's Energy Crisis: Reality or Hysteria?" *Bulletin of the Atomic Scientists,* 27, No. 7 (September 1971), p. 3.

Squires, Arthur M. "Clean Power from Coal." *Science,* 169, No. 3948 (August 28, 1970), pp. 821-27.

Starr, Chauncey. "Energy and Power." *Scientific American,* 224, No. 3 (September 1971), pp. 36-50.

Summers, Claude M. "The Conversion of Energy." *Scientific American,* 224, No. 3 (September 1971), pp. 148-64.

Tamplin, Arthur R. "Issues in the Radiation Controversy." *Bulletin of the Atomic Scientists,* 27, No. 7 (September 1971), p. 25.

Thienes, C. H., *et al.* "Effects of Ozone on Experimental Tuberculosis and on Natural Pulmonary Infections in Mice." *American Industrial Hygiene Association Journal,* 26 (May-June 1965), pp. 255-60.

Thompson, C. Ray, and Ivie, J. O. "Methods for Reducing Ozone and/or Introducing Controlled Levels of Hydrogen Fluoride Into Airstreams." *International Journal of Air Water Pollution.* London: Pergamon Press, 1965, pp. 799-805.

Thorp, Clark E. *Bibliography of Ozone Technology.* Chicago: Armour Research Foundation of Illinois Institute of Technology, Vol. 1 (1954), Vol. 2 (1955).

———. "Ozone." *Standard Handbook for Electrical Engineers,* 9th ed., New York: McGraw-Hill, 1957, pp. 1691-92.

Tranen, J. D., and Wilson, G. L. "Electrostatically Induced Voltages and

Currents on Conducting Objects under EHV Transmission Lines." *IEEE Transactions on Power Apparatus and Systems,* PAS-90, No. 2 (March-April 1971), pp. 768-77.

Underground Power Transmission. Report to the Federal Power Commission by the Commission's Advisory Committee on Underground Transmission, April 1966.

Vassell, Gregory S., and Maliszewski, Raymond M. "AEP 765-kv System: System Planning Considerations." *IEEE Transactions on Power Apparatus and Systems,* PAS-88, No. 9 (September 1969), pp. 1320-28.

Wehner, D. M. D. "Electro-Aerosols, Air Ions and Physical Medicine." *American Journal of Physical Medicine,* 48, No. 3 (June 1969), pp. 109-50.

Werthamer, S.; Schwartz, L. H.; Carr, J. J.; and Soskind, L. "Ozone-Induced Pulmonary Lesions." *Respiratory Care,* 16, No. 1 (January-February 1971), Abstracts, pp. 18-19.

White, Harry J. *Industrial Electrostatic Precipitation.* Reading, Mass.: Addison-Wesley, 1963.

White, Harry J., and Cole, William. "Design and Performance Characteristics of High-Velocity, High-Efficiency Air Cleaning Precipitators." *Journal of the Air Pollution Control Association,* 10, No. 3 (June 1960), pp. 239-44.

White, Irwin L. "Energy Policy-Making." *Bulletin of the Atomic Scientists,* 27, No. 8 (October 1971), pp. 20-26.

Woodwell, George M. "The Energy Cycle of the Biosphere." *Scientific American,* 223, No. 3 (September 1970), p. 64.

Young, F. S. "Underground Transmission Prospective." 1970 Electric Utility Engineering Conference, March 15-27, 1970, Subject No. 58.